Ministry of Agriculture, Fisheries and Food

Report of the Committee to Consider the Ethical Implications of Emerging Technologies in the Breeding of Farm Animals

London: HMSO

The Rt Hon William Waldegrave MP
Minister of Agriculture, Fisheries and Food

December 1994

Dear Minister

I have pleasure in presenting to you the Report of the Committee on Ethical Implications of Emerging Technologies in the Breeding of Farm Animals.

If any member of the Committee was previously unaware of the degree of public interest in matters we have examined, the response to our consultation letter revealed to us all the depth of concern surrounding these issues. The views expressed to us were, of course, extremely varied and the composition of the Committee was itself intended to reflect a diversity of background and approach. We have found it possible, however, to reach agreement on the course which future policy should take, and our recommendations to you are unanimous.

The Report has three major elements. In the first place we have made a general recommendation relating to the principles which should govern the treatment of animals. The fact that the development of the emerging technologies is not finished, but is very much still in progress, makes it important to set out these principles. We believe that future policy in this area needs such principles to guide it, and also that the appointment of a standing committee with a watching brief in this and other areas is a matter of some importance. In the second place we have made a series of more specific recommendations in relation to the application of some of the technologies, and in the light of our general principles, to ensure the proper treatment of the animals involved. And then, in the third place, we have reviewed the existing regulations relating to other areas of concern, commented on their adequacy and made recommendations accordingly.

We believe that the United Kingdom has taken a lead in the regulation of the emerging technologies in general and of genetic modification in particular, and that continuing to do so is vital to the maintenance of public confidence in farming as it enters the new world which is being opened up by scientific advances. We commend our Report to you as providing a framework for ethical discrimination and thus the basis for regulation not only of present applications of these scientific advances, but also of those which future research may yet render possible.

Yours sincerely

Michael Banner
Chairman

ACKNOWLEDGEMENTS

I am grateful to my colleagues who have served with me on the Committee for the time they have given to our deliberations and for the co-operative and friendly spirit by which our discussions have been animated. I must also express the gratitude of the entire Committee to our secretariat, Mr Stewart Marshall, Miss Kate Ward and Mr Paul Kilby, for the ready and invaluable assistance they have given us.

Michael Banner
Chairman

COMMITTEE TO CONSIDER THE ETHICAL IMPLICATIONS OF EMERGING TECHNOLOGIES IN THE BREEDING OF FARM ANIMALS

Chairman

The Reverend Professor Michael Banner MA, DPhil
F.D. Maurice Professor of Moral and Social Theology, King's College, London.

Members

Professor Grahame Bulfield BSc, DipAnGen, PhD, FRSE

Director, Roslin Institute (Edinburgh). Honorary Professor of Genetics and Agriculture at the University of Edinburgh.

Professor Stephen Clark MA, DPhil

Professor of Philosophy at Liverpool University.

Mr Luke Gormally LicPhil

Director of the Linacre Centre for Health Care Ethics and member of the Catholic Bishops' Joint Committee on Bioethical Issues and of its Standing Committee.

Mr Peter Hignett MRCVS

Retired veterinary surgeon, who during his career had particular interest in reproductive problems in animals, particularly cattle. Chairman of the Disciplinary Committee, Royal College of Veterinary Surgeons.

Mrs Harriet Kimbell LLB, LLM

Solicitor and lecturer at the College of Law. Deputy Chairman of the Council of the Consumer's Association, member of the Consumer Panel and the Polkinghorne Committee on the Ethics of Genetic Modification and Food Use.

Miss Cindy Milburn BSc, MSc

Special Projects Director for the World Society for the Protection of Animals, and member of the Farm Animal Welfare Council, and the Genetics Forum.

Mr John Moffitt CBE, DCL, FRASE

Dairy farmer in Northumberland. Member of advisory committees at the BBSRC Institutes at Babraham (Cambridge) and Roslin (Edinburgh), and the Centre of Genome Research.

Secretariat

Mr Stewart Marshall
Miss Kate Ward
Mr Paul Kilby

CONTENTS

SUMMARY AND LIST OF MAIN RECOMMENDATIONS

1. By our terms of reference we were invited 'to consider the ethical implications of emerging technologies in the breeding of farm animals; to advise on the adequacy of the existing legal and other safeguards in those areas; and to make recommendations'. We proceed by identifying and considering six main areas of concern.

2. In Chapter 2 we set out an ethical framework in the light of which we address these concerns. The main element in this framework is a set of principles which properly govern the treatment of animals and which express the view that the use of animals is permissible providing that use is humane. The humane use of animals respects the following three principles.

> (a) Harms of a certain degree and kind ought under no circumstances to be inflicted on an animal.

> (b) Any harm to an animal, even if not absolutely impermissible, nonetheless requires justification and must be outweighed by the good which is realistically sought in so treating it.

> (c) Any harm which is justified by the second principle ought, however, to be minimized as far as is reasonably possible.

We maintain that these principles provide a proper basis on which to evaluate the emerging technologies and draw particular attention to the fact that the first principle is the starting point of an adequate system of regulation. It follows that we regard a cost/benefit approach (which weighs the harms to the animals against the benefits from their use) as insufficient on its own since it fails to take account of the fact that some uses of animals are objectionable in themselves, regardless of any further consideration of their overall consequences. It also follows from our adopting the above principles that we do not accept the claims of those who, on the one side, reject any use of animals for human benefit and also of those, on the other side, who regard any use of animals as acceptable and are opposed to any restraint on farming practice. We take the view that it is the task of the Government to establish a framework in which animals are treated humanely, and that the application of our three principles would achieve that end. We do not rule out that in establishing such a framework, the Government may properly act independently of any European or international consensus on these matters.

3. In Chapter 3 we apply the first of our three principles and ask whether the application of the emerging technologies is intrinsically objectionable on the grounds that they fail to respect the natural characteristics, dignity and worth of animals. We argue that some possible uses of the new technologies are objectionable in these terms – for example, possible uses of genetic modification in which animals are treated as raw material upon which our ends and purposes can be imposed regardless of the ends and purposes which are natural to them.

4. We apply the other two principles in Chapter 4 and identify two further classes of uses of the new technologies – uses which are not intrinsically objectionable and absolutely impermissible, but which are only justified in particular circumstances where a substantial good is expected; and uses which are in general acceptable, but in which, nonetheless, all reasonable steps should be taken to ensure that any harms to the animals are minimized.

5. Our conclusions and recommendations in Chapters 3 and 4 fall under the following three classes.

(a) In relation to those uses of the new technology we have judged intrinsically objectionable – in particular those instances of genetic modification which can be thought to constitute an attack on an animal's essential nature – we consider whether the relevant current regulations (i.e. the Animal (Scientific Procedures) Act 1986 (ASPA)) are adequate to prohibit such modifications. Our view is that these regulations are broadly speaking adequate, but that clarification is required on two points – the interpretation of the key notion of 'adverse effects' and the means by which the existence of such effects is established. We note, however, that in a rapidly changing field there is a need for a standing committee in which the ethical questions which future developments will pose can be rigorously addressed. We believe that such a standing committee would play an important role in creating greater trust between industry, scientists and the public.

(b) Secondly, in relation to uses of the emerging technologies which are not absolutely impermissible but are justified only in particular circumstances – such as non-intravaginal AI in sheep and goats, and embryo transfer in sheep, goats, pigs and deer, which all involve non-therapeutic surgery – we recommend a variety of regulations which will effectively prohibit their routine use.

(c) Thirdly, where we judge certain uses of the emerging technologies to be generally acceptable – such as artificial insemination (AI) in cattle and pigs – we have made recommendations to ensure that any harms caused are, within reason, minimized.

6. In Chapter 5 we consider whether there are ethical grounds for denying protection by patent to work relating to the genetic modification of animals. Such work may be protected by patents relating to the gene construct, the processes of modification, the techniques for inserting modified genes or the modified animal itself. Some members of the Committee take the view that patents should not be granted on the modified animal – for one reason because the granting of such patents makes sense only if the animals are to be regarded as inventions. The majority takes the view that the granting of a patent on a particular form of an animal need not be understood in this way and thus need not be considered to be objectionable, since the patent relates only to the animal as expressing a specific form which is undoubtedly the product of human ingenuity and does not relate to the animal as such. Regardless of our difference on this particular question, we all accept that the case for denying protection by patents to *all* work relating to genetic modification of animals is not finally persuasive. We believe, however, that a moral criterion does have a proper place in the consideration of patent applications. The European Union (EU) draft Directive on the Legal Protection of Biotechnological Inventions has such a criterion and denies patents to processes which cause more harm to animals than benefits to mankind or other animals. In line with our stated principles, we would prefer a stronger moral criterion – specifically one that ensured that patents would not be granted where the resultant animal is such that its natural good or integrity has not been respected by the modification. However, in spite of reservations about the Directive, we believe that it can be supported insofar as it goes some way towards protecting animals from unwarranted genetic modification.

7. In Chapter 6 we argue that though the assessment of the nature and degree of risk which the release of genetically modified organisms may pose to the environment is a scientific matter, the question as to what risks are tolerable is an ethical question. It is a question, in other words, about what one values and to what degree, and thus cannot be answered by science. The regulations which are currently in place are designed to ensure that genetically modified organisms are securely contained and that where consent is sought for their release, the risks are carefully assessed and consent given only if there is no real threat of harm to the environment. We note in particular that the Secretary of State is under a duty to prevent or minimize any damage to the environment and is bound, therefore, to decline consent for any release which poses a serious risk to the environment. We endorse the cautious, case by case approach which the regulations enshrine and believe that, properly enforced, they offer an appropriately high level of protection to the environment without placing improper constraints on industry.

8. In Chapter 7 we argue that there is no reason to suppose that the new breeding technologies will necessarily have an impact on genetic diversity in farm animals. The interest of breeders lies in maintaining variation, and the increasing use of these technologies will not alter that situation nor discourage breeders from acting so as to protect diversity. The loss of rare breeds, which may or may not mean a loss of valuable genetic variation, has been caused by many factors, and cannot be blamed on the new technologies; indeed these new technologies may have a part to play in assisting in the preservation of rare breeds by enabling the effective use of the remaining animals in breeding programmes which will ensure their survival. Nonetheless we believe that further consideration should be given to the need for specific measures to preserve rare farm breeds.

9. In Chapter 8 we discuss concerns surrounding the socio-economic consequences of the new technologies. We acknowledge that there is considerable difficulty in anticipating the socio-economic effects of the introduction of the emerging technologies on the levels of production, the size of farms and the position of the Third World, and conclude that until trends become clear the situation can only be monitored. We take the view, however, that one unlooked-for consequence of the introduction of the emerging technologies can quite reasonably be anticipated – and that is the creation of public suspicion of farming, unless those who are engaged in the development and application of these technologies endeavour to be sensitive to public concerns, open to debate with interested parties and supportive of a reasonable system of regulation, provision of information and labelling.

List of recommendations

Our recommendations are as follows:

General principles

(1)   That the following principles be accepted as a framework within which present and future uses of animals should be assessed.

   (a) Harms of a certain degree and kind ought under no circumstances to be inflicted on an animal.

   (b) Any harm to an animal, even if not absolutely impermissible, nonetheless requires justification and must be outweighed by the good which is realistically sought in so treating it.

   (c) Any harm which is justified by the second principle ought, however, to be minimized as far as is reasonably possible (para. 2.18).

Intrinsic objections

(2)   That an advisory standing committee be created, whose remit should include a responsibility for broad ethical questions relating to current and future developments in the use of animals (para. 3.34);

Breeding techniques and welfare

(3) That in the review of the approval procedure for non-veterinarian inseminators of cattle, there should be no relaxation of the requirement that non-veterinarians seeking to carry out this work should have reached a high level of competence (para. 4.26);

(4) That non-veterinarians carrying out intravaginal artificial insemination should be required to hold a certificate of competence from an appropriate body (para. 4.31);

(5) That non-intravaginal artificial insemination should continue to be performed only by veterinary surgeons competent in the procedures and that, in consultation with the Farm Animal Welfare Council (FAWC) and the Royal College of Veterinary Surgeons (RCVS) in particular, the following provisions be introduced by code of practice or, where appropriate, by regulations under the Animal Health and Welfare Act 1984, to regulate non-intravaginal AI in sheep and goats, unless carried out under ASPA:

(a) that laparoscopic AI be performed with appropriate and adequate analgesia;

(b) that in all cases of non-intravaginal insemination, a veterinary surgeon should review the health, maturity and general suitability of the animal to be inseminated in order to ensure, as far as possible, a normal pregnancy and delivery;

(c) that non-intravaginal AI be used only in disease control programmes and in recognised breed improvement schemes (para. 4.33);

(6) In relation to AI centres for pigs, that the system of licensing should continue, and that non-veterinarian practitioners of pig AI should be required to hold an approval certificate to be obtained from an appropriate body (para. 4.40);

(7) That embryo transfer in sheep, goats, pigs and deer should only be performed by veterinary surgeons competent in the procedure and that, in consultation with FAWC and the RCVS in particular, the following provisions be introduced, by code of practice or where appropriate regulations under the Animal Health and Wefare Act 1984, to regulate embryo transfer in these species, unless carried out under ASPA:

(a) that embryo transfer be performed with appropriate and adequate analgesia;

(b) that in all cases a veterinary surgeon should review the health, maturity and general suitability of the animal to receive an embryo, to ensure as far as possible, a normal pregnancy and delivery;

(c) that embryo transfer in these species be used only in disease control programmes and in recognised breed improvement schemes (paras. 4.60, 4.64 and 4.66);

(8) That if per vaginam methods of embryo transfer for pigs are developed, an approval procedure should be established to ensure that technicians seeking to carry out this work should do so under veterinary supervision and should have reached a high level of competence before they receive a licence (para. 4.65);

(9) That regulations be made under the Animal Health and Welfare Act 1984 requiring that non-veterinarians should carry out embryo transfer in horses under veterinary supervision and only if they hold an approval certificate granted by the appropriate body after suitable training (para. 4.67);

(10) That FAWC review the evidence on ultrasound scanning in bovines and, if necessary, regulations be made, as appropriate (para. 4.71);

(11) That FAWC should review the issue of the use of ovum pickup (para. 4.76);

(12) That the Animal Procedures Committee be invited to give an account of how the existence of 'adverse effects' is established and to address whether genetic modifications which could be judged intrinsically objectionable would be held to have caused 'adverse effects' under ASPA (para. 4.104);

Patenting

(13) That developments should be monitored in relation to the threat to small producers posed by widely drawn claims to patent protection (para. 5.22);

(14) That the draft EU Directive on the legal protection of biotechnological inventions should be supported as it relates to animals (para. 5.31);

Genetic modification and environmental risks

(15) That the Government continue to support international understanding, harmonization and co-operation on the control of genetically modified organisms (para. 6.25);

(16) That the Advisory Committee on Release into the Environment (ACRE) should continue to scrutinize applications for release or marketing of genetically modified organisms on a case-by-case basis and impose restrictive conditions until research or experience has provided sufficient data on the impact of releases to allow any relaxation of those conditions (para. 6.27);

Impact on genetic diversity

(17) That further consideration be given, by Government, to the need for specific measures to conserve farm animal breeds. Such measures, whether appropriately sponsored by Government or by others, might include for example:

(a) the establishment of a UK register of breeds, to record their numbers and population sizes;

(b) a survey to measure diversity within and between breeds using molecular markers and production traits;

(c) the construction of a biodiversity database;

(d) the establishment of a genome bank where gametes and embryos are cryogenically stored for use at a later date to re-introduce genes that have been lost from a population (para. 7.25).

CHAPTER 1:   INTRODUCTION

Terms of reference and background

1.1   In April 1993 Ministers appointed a committee with the following terms of reference: 'to consider the ethical implications of emerging technologies in the breeding of farm animals; to advise on the adequacy of the existing legal and other safeguards in those areas; and to make recommendations'.

1.2   The use of **selective breeding**[1] to improve the quality of farm animals for the various purposes for which they are kept has been rendered more effective in the last hundred years or so by the growth of the science of genetics, but further progress has been made by the use of a range of technologies, some of which are already well-established, others of which have, as yet, no commercial application.

1.3   The emerging technologies which we have been asked to consider are a family of techniques which, broadly speaking, regulate the reproduction of animals. These techniques are described in later chapters of this report, but here it will be sufficient to place them in three main categories. In the first place, there are those which enable the more effective use of the male in a breeding programme: **artificial insemination** has been in widespread use since the 1930s and has contributed significantly to the performance of farm animals. In the second place, there are techniques which allow the more effective use of the female element: **superovulation**, **embryo transfer** and *in vitro* **fertilization** enable many more eggs from a preferred animal to be fertilized and brought to term. And then in the third place, there are those techniques which are aimed not at the wider use of favoured **gametes**, but at increasing the number of particular animals with the required characteristics (through **cloning**), or altering the characteristics of animals (through **genetic modification**). Taken together, these represent a powerful range of technologies which supplement and go beyond what can be achieved by conventional breeding and open up many possibilities for the farming of the future.

1.4   By our terms of reference we are invited to consider the ethical questions which such new technologies raise and the adequacy of existing legal and other safeguards. This is a timely request in a number of ways, but chiefly because it is opportune, while some of these techniques are at the development stage, to review the present system of regulation and to ask whether it is adequate not only for now but for the future. Public confidence in, and acceptance of, the application of the emerging technologies is quite properly dependent upon its confidence in a system of regulation which ensures the ethical acceptability of developments in this field. Without this confidence, suspicion may attach to developments which are in actual fact perfectly benign and may have the potential to play an important role in meeting pressing and legitimate needs for food, pharmaceuticals or other products. Thus a system of regulation which is not, or is not perceived to be, adequate to the task, may itself hamper the development of the new technologies by contributing to public unease. We believe, therefore, that all those who have an interest in this field will welcome the opportunity to take stock of the diverse measures which currently regulate this area and to consider their suitability in the face of current and possible future developments.

1.5   Many of the technologies we shall consider can be used in the manipulation of human reproduction. The many questions which such practices raise are outside our remit. We have also not dealt with the ethical questions raised by the consumption of

---

[1] Technical terms are shown in bold when used for the first time and described in the Glossary at Annex D.

genetically modified foods. These questions have been addressed by the *Report of the Committee on the Ethics of Genetic Modification and Food Use*[2].

Method of work

1.6 The Committee began by issuing a consultation letter (Annex A) in which we identified the following areas as being of ethical concern and as calling for consideration:

> (a) intrinsic objections to the use of the emerging technologies in general and genetic modification in particular;

> (b) the effect of the various techniques on the welfare of farm animals used for breeding purposes and on their progeny;

> (c) the appropriateness of the use of patent law to protect commercial developments in relation to genetic modification;

> (d) the risks, if any, to the environment which may arise from novel breeding technologies;

> (e) the effect of these technologies on the genetic diversity of farm animals;

> (f) the impact of the use of new technologies on the social and economic life of the community as a whole.

1.7 We invited written submissions from a large number of organisations (Annex B) on these and on any other matters of concern which came within our terms of reference. We were very grateful for the many responses we received; a list of respondents is at Annex C. The Committee met for the first time in October 1993 and on seven other occasions. Visits have been made to Genus in Northumberland and to the Roslin Institute in Edinburgh, and a small number of expert witnesses were invited to attend meetings to answer questions and to enable the Committee to hear their views.

---

[2] The *Report of the Committee on the Ethics of Genetic Modification and Food Use* was published by HMSO in 1993 (ISBN 0–11–242954–8).

CHAPTER 2:   GENERAL PRINCIPLES

2.1   It is appropriate at the outset to say something of the ethical approach which we have adopted in what follows. Our task has been to consider ethical concerns which arise in relation to the emerging technologies which have been referred to in the previous chapter. But within what ethical framework has that consideration taken place? After all, if we focus for the moment on questions which have naturally been our chief concern, namely those relating to the status and protection of animals, public views differ widely between those who regard any use of animals as morally acceptable, and others who regard all uses as morally unacceptable.

2.2   We have not thought it appropriate or necessary to begin by arguing directly with either of these widely differing views, but have approached our task by considering the adequacy of the general principles which seem to underlie the present regulations governing the treatment of animals, to see whether they can properly be applied to the problems before us.

2.3   As we understand them, current regulations which govern the use of animals are based on and express the broad principle that such use of animals, for any purpose, agricultural or otherwise, is acceptable, provided the use is humane. As it stands, of course, this principle is extremely general and its further definition and application is a matter of controversy. However, even in its general form it is important since it represents the culmination of a long tradition of moral reflection, as well as expressing the views of most members of society, that the use of animals is, morally speaking, neither absolutely impermissible, nor a matter about which one should be indifferent.

2.4   This broad principle is made more specific in the current regulations relating to the use of animals by the following three more detailed principles.

   (a) Harms of a certain degree and kind ought under no circumstances to be inflicted on an animal.

   (b) Any harm to an animal, even if not absolutely impermissible, nonetheless requires justification and must be outweighed by the good which is realistically sought in so treating it.

   (c) Any harm which is justified by the second principle ought, however, to be minimized as far as is reasonably possible.

2.5   The first principle provides the rationale for the prohibition of numerous non-therapeutic operations on farm animals – tongue amputation in calves, tail docking in cattle and toothgrinding in sheep, for example. These regulations are based on the principle that certain harms caused to animals should have no place in farming practice.

2.6   The second principle is implicit in the Animals (Scientific Procedures) Act 1986 (ASPA), for example, and ensures that animals are used in experimental work only where the end result of the experiment can reasonably be expected to be commensurate with the harm suffered by the animals. The same principle should govern the use of farm animals, since, no less than laboratory animals, they ought to be protected against harms which lack any adequate overall justification. It should be stressed that this principle enshrines a necessary but not sufficient test of the moral acceptability of any use of an animal, since – following the first principle – harms of a certain degree and kind ought not to be inflicted on an animal in any circumstances.

2.7   The third principle is implicit in a large number of codes which, while accepting that certain sorts of procedures involving harm to animals are, in general, acceptable, nonetheless seek to ensure that the harms caused are minimized by good practice. A recent example of a code expressing this principle is the Bovine Embryo Collection and Transfer Regulations 1993, which we will have cause to refer to in more detail in Chapter 4.

2.8 It may be important to add for clarification, that by the word 'harm', employed in these principles, we do not mean to refer only to harm of which the animal is conscious or even simply to physical harm. We would contend that animals can be harmed or wronged in other ways than simply by physical mistreatment. An animal can be harmed, for example, by treatment which is degrading. Thus one might object to the dressing of animals in human clothes for public spectacle, even though the animal so treated may be neither conscious of any wrong being done to it, nor the object of physical mistreatment.

2.9 Taking 'harm' in this wide sense, the three principles we have stated express a defensible and coherent approach to the regulation of the agricultural use of animals, and have informed our consideration of the issues relating to animal welfare which arise from our review of the emerging technologies. We recognize, of course, and have made clear in what follows, that the application of these principles is by no means easy in real cases, but we believe, nonetheless, that these principles provide the foundation for the proper regulation of the emerging technologies we have been asked to consider. The application of each of these technologies, in so far as it causes harm to animals, must be subject to three tests. Is the harm of such a kind that it is objectionable and impermissible? If the harm is not of this kind, is the harm warranted by the good which the technique promises? And if the harm is so warranted is it minimized, so far as is reasonably possible, by good standards of practice?

2.10 We draw particular attention to the fact that the first principle has an important place in the consideration of the treatment of animals. There has been a tendency in some approaches to these issues, and in some of the submissions we have received, to apply a so-called 'cost/benefit analysis' of the harm done to animals as a sufficient test of the moral acceptability of what is under review. There is no doubt that for all the real difficulties with such an analysis, the need to weigh the good and bad consequences which can be expected to flow from a particular course of action encourages a careful review of what is at stake and is appropriate as a means of protecting animals, whether in laboratories or on farms, from unwarranted procedures. It cannot function, however, as the sole test of the acceptability of particular uses of animals but must be augmented by a consideration of whether the action which is proposed, either in itself or in virtue of its particular consequences, ought not to be done.

2.11 The fact that we regard these two tests as necessary, and not just the cost/benefit analysis alone, indicates our disagreement with the approach to moral questions known as consequentialism. Since this approach is not without influence, we have felt obliged to explain why we do not consider it satisfactory and we have done so in Chapter 3.

2.12 The constraints which we consider are appropriately placed on our use of animals would be judged insufficient by those who advocate 'animal rights' and take the view that animals (which is to say non-human animals) should be given something close to the same respect which is given to human animals. From this perspective it is maintained that any creature with a life of its own, especially one whose behaviour is so sophisticated as to suggest that it possesses preferences and intentions, should not only be treated respectfully, but should be accorded something like the rights to life, liberty and the pursuit of happiness which we accord members of our own species. In the light of this standard, almost all current agricultural and scientific uses of animals will be judged wrong; thus, whereas animal welfarists aim to prevent or reduce the amount of animal suffering associated with our use of animals, those who argue for animal rights take exception to the use of animals as such and to what they would regard as our enslavement and murder of them.

2.13 It follows that for those who believe in animal rights (and some members of the Committee are sympathetic to the animal rights position) the recommendations of this Committee, which arise from the majority viewpoint, constitute no more than palliatives to a system which is fundamentally unacceptable. The view of the majority of the Committee is that the use of animals is acceptable, provided that use is humane, and our recommendations are aimed at securing the more humane use of animals. Advocates of animal rights will disagree with the underlying approach we have adopted, but ought, nonetheless, to be able to welcome the Committee's proposals – whilst they take the view that animals should be accorded something like human rights, they can perhaps support measures to ensure they are treated humanely.

2.14 If there are some who will consider our recommendations insufficiently radical, those same recommendations will doubtless be criticized from the other side by those who are opposed to any restraints on the use of animals in farming. Whilst very few actually espouse this extreme view, there are, it seems, a good many who think any new provision aimed at protecting farm animals is unacceptable if it may in any way threaten the profitability of British farms. Such provisions are only acceptable, so it is said, if they are agreed at the European level – if they are not agreed at this level, it is contended, we are simply disadvantaging our own farmers and 'exporting our welfare problems', since farmers on the Continent will now produce for the British market what British farmers are prohibited from producing. This argument is so frequently stated, and stated as if it is hardly open to question, that we think it necessary to comment on it.

2.15 Whilst it is obviously highly desirable that provisions relating to farm animal welfare be adopted in Europe, and indeed as widely as possible, we do not accept the contention that the UK Government should in all cases decline to take action prior to agreement at the European level. Doubtless abolition of child labour in nineteenth century Britain could have been opposed on the grounds that it would have disadvantaged British manufacturers and simply result in the export of our child welfare problems, but neither contention should have been found persuasive. Of course moral conduct can be costly, but it can hardly be argued that we should delay behaving properly until we can guarantee that so behaving will cost us nothing. After all, if we are excused from following our principles until there is European agreement, why not until there is world-wide agreement? Nor does the point about 'exporting our welfare problems' carry much weight. The fact that someone else is prepared to do something which we judge to be wrong and which we consequently decline to do, is certainly regrettable – but it does not usually persuade us to behave wrongly in the first place. If a car is left unattended with its keys in, it is almost certain to be stolen, but if one were to steal it one should not expect to be excused blame by pleading that someone else would have done it in any case.

2.16 The task of the Government is to establish appropriate protection for farm animals in Britain whether or not that protection commends itself to our European partners. In any case we do not believe that unilateral action in relation to animal welfare will necessarily threaten the economic position of British farms. There are two main points to be made. First of all, measures to protect animals are not invariably burdensome – some may cost nothing or so little as to be economically insignificant. But the second point is that it is distinctly unimaginative to suppose that the establishment of a proper framework for the protection of the well-being of farm animals poses a threat, and not an opportunity, for British producers. At a time when the consumer is increasingly conscious of farm animal welfare, the fact that British produce is produced in accordance with the highest welfare standards offers a marketing possibility of potentially great value. Having said that, however, we recognize that in some cases there may be costs in acting unilaterally on welfare matters, but as we have already observed, the claims made upon us by ethical principles cannot simply be suspended where those principles seem to conflict with self-interest.

2.17 Some of the concerns relating to these technologies do not have to do with issues regarding the status and protection of animals – for example, concerns about the impact of the use of the emerging technologies on genetic diversity, the environment and the socio-economic position. Nonetheless, in addressing these problems we have followed the same ethical framework which we have outlined in this and the next chapter.

Conclusion

2.18 The humane use of animals respects, we maintain, three principles (para. 2.4). We recommend that these principles be accepted as a framework within which present and future uses of animals should be assessed.

CHAPTER 3: INTRINSIC OBJECTIONS

What are intrinsic objections?

3.1 In our consultation letter we listed intrinsic objections to the use of the new technologies as one of the issues which we proposed to consider. What did we mean to refer to by this phrase?

3.2 An intrinsic objection to a particular practice or action is an objection which does not relate to the practice's consequences or effects, but to the practice or action itself. Thus when we complain of some action that it is dishonest, deceitful or disloyal, for example, we are not necessarily objecting to the effects of the action – these may be good or bad, depending on the circumstances – but to the action itself. The action is dishonest, deceitful or disloyal in virtue of the sort of action it is, and would thus be judged objectionable. This is not to say that all actions which are judged intrinsically objectionable ought never to be performed – someone might well hold that in certain circumstances, such as when loyalty would threaten a grave miscarriage of justice, disloyalty to a friend is permissible. But there are some intrinsically objectionable actions or practices which would usually be thought absolutely impermissible – the enslaving of one human being by another or the torturing of children would be examples.

Intrinsic objections to the emerging breeding techniques

3.3 It is clear from the responses we have received as well as from public discussions of this topic that, though they may not use this language, many people have intrinsic objections to the use of the emerging technologies. They may well be concerned about the effect of these technologies on animal welfare, genetic diversity, the environment, the pattern of farming and rural life, etc., but their concerns would not be exhausted by a consideration of these matters. For as well as worrying about the effects of the new technology, they feel a distinct unease about its very use.

3.4 Behind the expressions of misgivings in a number of the submissions made to the Committee lies the conviction that the application of the new technologies to farm animals involves an essentially improper attitude towards them, expressing, in effect, the view that animals are no more than raw material for our scientific projects or agricultural endeavours. The view that animals are no more than raw materials, it would be argued, fails to take account of the fact that the natural world in general, and animals in particular, are worthy of our respect as possessing an integrity or good of their own, which we ought not simply to disregard. To seek to manipulate what is given to us as if it were a formless lump of clay is fundamentally disrespectful and an expression of overweening human pride or hubris – it is, as it is sometimes dramatically put, an attempt to 'play God'.

3.5 It is important to stress that this objection might well be voiced by someone who is not against the use of animals altogether, nor against the use of animals for food in particular. Certainly it will be put by those who believe in animal rights and hold that any use of animals is an infringement of these rights; but it could be put by others who are content that we should keep animals for a variety of uses and purposes. They would contend that there are ways of using animals, albeit for human advantage, which nonetheless respect the natural characteristics and good of the animals, but that the radical moulding of animals to suit our purposes which the new technologies makes possible steps over a boundary, and is expressive of a disrespectful attitude.

3.6 The objection might be put from a Christian perspective, in which the world is regarded not simply as a chance outcome of diverse forces, but as a created order which has been shaped by God and as such is to be accepted by us as having a good or integrity of its own. Other theistic religions might advance the same objections – indeed representatives of a number of non-Christian faiths expressed concerns of this sort to the recent Committee on the Ethics of Genetic Modification and Food Use.

3.7 But the objection need not be grounded in an explicitly theological or religious doctrine: even those who do not regard the living world as a created order may feel a profound respect for living creatures, as individuals and species. Stephen Jay Gould, for example, refers to the "cascade of astounding improbability stretching back for millions of years"[3] which has produced the present pattern of life, and insists that biologists typically and properly respect the "integrity of nature". Individuals, species and the whole process of evolutionary change deserve respect, even if 'mere chance' engendered them.

3.8 It is interesting to note that very few of those who responded to our letter of consultation and who were, broadly speaking, well-disposed to the new technologies, actually addressed or answered this particular concern. Some seemed to think that if the effects of a technology are shown on balance to be good, there could be no reasoned opposition to it. Hence, they assume that any opposition which does not focus on the supposed ill-effects of the use of new technologies can only be explained by the disparagingly-labelled 'yuk factor': an emotional reaction to the introduction of a technology which is quite without rational warrant, and can be expected to disappear as people become accustomed to that technology.

3.9 In our view the intrinsic objection to the use of emerging technologies which we have stated, from whatever perspective it is put, makes an important point, is not to be treated lightly and cannot be discounted. Certainly the fact that the objection is often stated in emotional terms is not sufficient reason for discounting it: revulsion or disgust at certain uses of animals may be perfectly rational and founded upon a conviction, of the sort we have tried to explain, as to the intrinsic wrongness of what is proposed. It is important, then, that we should give careful consideration to the contention that the application of the emerging technologies to farm animals is intrinsically objectionable.

Philosophical doubts about intrinsic objections

3.10 Before considering that objection further, however, it is necessary to address directly those who think that there are philosophical grounds for discounting the very possibility of intrinsic objections. In taking seriously the notion that the application of the new technologies to farm animals might be intrinsically objectionable we disagree with the assumptions involved in a so-called consequentialist approach to moral issues. Although they may not be aware of it, we suspect that some of those who are quick to discount intrinsic objections have fallen under the influence of this philosophy.

3.11 Consequentialism holds that actions are good or bad, right or wrong, solely in virtue of their overall consequences. If that were true, it would not make sense to say of some action or practice that it is intrinsically objectionable, let alone absolutely impermissible – an action could only be judged wrong in the light of a consideration of its overall consequences.

3.12 It is taken for granted in our moral life, however, that some actions are intrinsically objectionable and that some of these intrinsically objectionable actions are absolutely impermissible: 'the ends do not justify the means' we say, holding that no matter the good results which may flow from particular actions, they cannot be

---

[3] Stephen Jay Gould, *Discover*, **6**(1), pp. 34–42, January 1985.

rendered acceptable. We have already given as examples of such actions the enslaving of one human being by another or the torturing of children. These are not the sort of actions which can be justified by a consideration of the balance of consequences which results from them: they are both intrinsically wrong and impermissible.

3.13 What is taken for granted in our moral life is not unassailable and perhaps consequentialism is right, someone might say, to challenge the presupposition that there are such things as intrinsically objectionable actions. This is not the place to conduct a detailed philosophical argument, but it is appropriate to point out that as a moral theory consequentialism has been subject to very severe and sustained philosophical criticism – so much so that it can hardly be regarded as a sound or acceptable basis on which to advance recommendations of public policy. In any case, the public to whom this policy must be acceptable believe certain things to be intrinsically objectionable, and some of them quite rightly wonder whether the application of the emerging breeding techniques to farm animals is one of those things. It would be unsatisfactory to rule this question out of court on the basis of an extremely dubious philosophical theory. Instead it must be addressed directly.

Consideration of the intrinsic objection

3.14 Some hold, as we have said, that the application of the emerging technologies to farm animals involves our taking an improper attitude towards them, treating them as little more than artefacts. We take this contention seriously, and recognize that there is a danger that the increasing element of technological intervention in the breeding of farm animals may contribute to the creation or reinforcing of a mentality which regards them as no more than commodities. However, we are not finally persuaded that the attitudes complained of are necessarily expressed in *all* applications of these new techniques. Certain uses of the techniques would indeed be objectionable for the reasons given, but others would be perfectly compatible with a proper respect for animals.

3.15 To delineate the distinction we seek to draw so sharply that any conceivable case is seen plainly to fall on one side of the line or the other may not be possible, but there are examples, we believe, which will serve to indicate the difference to which we intend to draw attention. It may be helpful to take for discussion a number of hypothetical cases which relate to genetic modification (for most people the most controversial of the new technologies), beginning with what we would regard as acceptable and ending with the unacceptable. Whether or not these modifications are presently technically feasible is immaterial for our purpose, which is simply to illustrate the type of genetic modification which might be thought intrinsically objectionable.

3.16 Suppose a scientist submits a proposal for three projects of genetic modification. The first is aimed at increasing the protein content of cows' milk with a view to increasing the milk's value to the human consumer. The second is aimed at causing poultry breeding stock to produce only female chicks to be raised as laying birds. And the third is aimed at increasing the efficiency of food conversion in pigs by reducing their sentience and responsiveness, thereby decreasing their levels of activity. Let us suppose in each case that the modification does not involve the use of genes from species other than the one to be modified – we shall give separate consideration to modification involving the transfer of genes between species.

3.17 The first of these proposed modifications is not, we would maintain, intrinsically objectionable. It seeks to enhance a particularly desirable trait just as traditional selective breeding does, and need not involve the attitudes complained of. Specifically the modification does not affect the animal's defining characteristics, nor threaten the achievement of its natural ends or good. Thus it does not treat the animal simply as a means to human profit or advantage, but respects its essential nature and well-being. We should stress that we are not saying that there will be no objections to what is proposed – certainly those who object to a system which removes calves from mothers at birth will object to this modification too, since it is designed to further what these critics regard as an unacceptable means of production. Our point is simply that those who accept the use of animals in general, and this system of production in particular, will find nothing new to take exception to in relation to the proposed modification itself.

3.18 The second case, where the modification aims to modify poultry breeding stock so that they produce only female chicks, is more radical in one sense, in that the end result cannot be thought of straightforwardly as 'enhancing a particularly desirable trait', even if it might be possible, by traditional means, to select birds which genetically produce more female than male offspring. It seems to us, however, that this difference is not morally significant. In terms of FAWC's 'five freedoms'[4], for example, this project would not deprive the chickens of the 'freedom to express normal behaviour' unless, rather implausibly, one included having offspring of two sexes in one's account of an animal's normal behaviour or well-being.

3.19 The third case is, however, different. Here the imagined modification aims to produce pigs of reduced sentience and disinclined to engage in the activity which is normal to them, thereby increasing the efficiency of their conversion of feed to meat. Even if this has no welfare implications (if welfare is understood narrowly as relating to an animal's happiness), so that by any available measure such pigs are as content as any other pigs, still we would maintain that the proposed modification is morally objectionable in treating the animals as raw materials upon which our ends and purposes can be imposed regardless of the ends and purposes which are natural to them. The fact that the project promises an increase in profit, or any other desirable consequence, does not, and cannot, wipe out the intrinsically objectional character of such an action.

3.20 Is genetic modification which transfers genetic material between species inherently objectionable as failing to respect animals in the forms or kinds in which they have been given to us? We would maintain that our assessment of the cases we have mentioned would not differ even if the modifications were to involve the use of genes from species other than the one to be modified; that is to say, that the key issue here is whether the modification respects an animal's natural characteristics or ends, and that such respect is possible even where genetic material is moved between species.

3.21 It is important, in the first place, to stress that species share a common genetic inheritance and that, further, the boundaries between them are not fixed but alter over time. The main point to be made, however, is that in most cases, the transfer of genetic material between species will not aim at the creation of new animals in any very interesting sense; that is to say, it will be aimed at the addition of a trait which does not alter in any significant way the animal's natural characteristics. For example, sheep have been bred at Edinburgh, which express a human protein in their milk as a result of the incorporation of a human **gene** in their **genotype**. But they are not sheep which have been rendered slightly human, but simply sheep. The underlying claim here is that it is a mistake to identify what is essential to an animal's natural existence as consisting in its present genetic make-up – only certain elements of an animal's genome are crucial in this respect. Of course, some changes in an animal's genotype could

---

[4] FAWC believes that the welfare of animals can be assessed by reference to 'five freedoms' which taken together define an ideal state for farm animals. The freedoms are: freedom from thirst, hunger and malnutrition; freedom from discomfort; freedom from pain, injury and disease; freedom to express normal behaviour; freedom from fear and distress.

constitute an attack on its natural existence – and these changes might be accomplished through genetic modification. There are, however, modifications, including modifications involving the use of genetic material from other species, which do not constitute such an attack. This is not to say that some of these modifications might not be objectionable for other reasons we have yet to consider; it is merely to say that some genetic modifications, even those involving a gene from another species, are not intrinsically objectionable in terms of the considerations we have so far outlined.

3.22 One particular use of genetic modification involving transfer between species has caused some public comment and is worthy of special mention as further explaining our approach. This is the attempt to generate a line of **transgenic** pigs which could provide organs and tissues suitable for transplantation into humans. The genetic modification might even be carried out with a specific patient in mind, thereby avoiding the problems of rejection which, along with the shortage of donor organs, currently hamper transplant surgery.

3.23 Given what we have said in relation to consequentialism, we do not regard it as a sufficient justification of this practice merely to point to the important benefits which it promises – some means of achieving undoubtedly good ends are both objectionable and impermissible. This use of genetic modification does not, however, fall into the category of uses of animals we have judged intrinsically objectionable. It does not treat the animals as raw material upon which our ends and purposes can be imposed *regardless* of the ends and purposes natural to them – supposing, that is, that their general well-being and characteristics are not affected by the alteration. This is not to say that there are no other moral questions raised by the project we have mentioned, but these do not fall within the scope of this Committee's terms of reference.

3.24 It is necessary to add immediately, however, that in this and the other examples we have discussed, we have assumed what could not be assumed in real cases, that these modifications have no adverse implications for the animal's welfare. Genetic modification is a science in its early stages, so that even if some modifications do not affect an animal's welfare, all attempts at modification risk causing real harm in this respect. Thus, whereas in case 1 and 2 what is aimed at may not be intrinsically objectionable, the risks involved in the attempt to achieve what is aimed at might render the attempt itself objectionable. That is to say, a fundamental disregard and contempt for animals could be expressed not only in seeking to override their natural forms, but in subjecting them to unwarranted risks of severe harm.

Discussion

3.25 The burden of this chapter is that intrinsic objections to the use of the new technologies cannot be lightly dismissed. Regardless of a consideration of overall consequences, there may be serious objections to breeding programmes making use of these techniques in that they fail to respect animals. It is not the case, however, that *all* uses of the new technologies are open to the objections we have considered. Some uses are intrinsically objectionable, others are not, and we have tried to distinguish, in broad terms and in relation to the issue of genetic modification, between the two. The key question here is whether a breeding programme threatens or respects an animal's natural characteristics and form.

3.26 If the only way of preventing the objectionable uses of the new technology were to ban any use whatsoever, it would be rational, though rather drastic, to recommend such a course. We take the view, however, that it ought to be possible to devise a policy which discourages unacceptable practices while allowing unacceptable uses of the new technology. Such discrimination is not only morally appropriate, but also has the advantage that it does not require us to forego the significant benefits which the emerging techniques promise. We further believe that the necessary structures for the implementation of such a policy are already in place. It will be most convenient to consider the policy we recommend in the light of the discussion of welfare issues which we undertake in the next chapter.

3.27 It is important to make the point here, however, that our argument has not been that the emerging technologies are morally indistinguishable from traditional methods, and since these are acceptable, so are the new techniques. If this had been our argument then it could quite properly be pointed out that sometimes the realization that our acceptance of a practice is logically implied or required by what we currently do causes us to reassess that current practice. Thus the contention that the principles we currently apply in relation to abortion oblige us in consistency to accept infanticide, might cause us to reconsider the principles which currently govern the practice of abortion rather than to legalize infanticide. Similarly, the contention that traditional farming practices are morally indistinguishable from those now emerging may cause us to realize just what is wrong with traditional farming practices. Our argument has not, however, consisted in a negative defence of these emerging technologies, but has rather rested on the contention that some uses of these technologies do not, as such, presuppose an improper disregard for the animals.

3.28 Far from taking as its foundation the acceptability of traditional farming practice, the argument we have developed commits us to a principle which may be as critical of certain programmes of selective breeding as it is of particular programmes employing the new technology. For example, any breeding programme which results in the impossibility of natural mating between normal members of a particular species would be objectionable, we would maintain, no matter the methods used in the programme. Although we do not regard AI or embryo transfer as impermissible, a breeding programme which was deliberately aimed at producing animals which could only breed by these means would constitute an improper attack on the essential nature of those animals and hence would be objectionable.

Conclusion

3.29 Those who object altogether to the use of animals will necessarily regard the genetic modifications we consider acceptable as quite unacceptable. We have approached this issue, however, in the light of the principles we have set out in Chapter 2. Thus we have tried to give sympathetic consideration to the concerns of those who, while not objecting to the use of animals as such, nonetheless regard the various new technologies, and in particular genetic modification, as straying over a boundary between acceptable and unacceptable treatment of animals specifically in using them as mere raw materials in disregard of their nature, worth and ends. We have concluded that some possible uses of the new technology would indeed be objectionable in this sense, and that for this reason, the tendency to regard a cost/benefit approach to these issues as in itself sufficient is mistaken. Since some uses of the new technology would be objectionable in themselves, regardless of any further consideration of their overall consequences, an adequate system of regulation must take as its starting point the principle we set out in the previous chapter: harms of a certain degree and kind ought under no circumstances to be inflicted on an animal.

3.30 In the next chapter we consider the adequacy of the current regulations governing genetic modification to prohibit objectionable uses of these particular techniques, but we are aware of the fact that it is not only genetic modification which can be objectionable on these grounds – conventional breeding programmes as well as others of the emerging technologies may be guilty of treating animals in such a way as to disregard their nature or dignity. We are also aware that there is presently no forum in which the intrinsic objections which, we believe, deserve a careful hearing, can be fully aired and debated. The brief of FAWC (para. 4.8) is limited, as its name indicates, to questions of welfare and it is chiefly concerned to monitor current farming practice; the Animal Procedures Committee[5] has a responsibility only in relation to scientific experimentation. We are further conscious of the fact that the technologies we have considered are developing very rapidly, and that applications of them which we have not envisaged, let alone addressed, may be just around the corner.

3.31 In view of the rapid scientific advances in this area and the consequent continual changes in the impact of the new technologies, we believe that there is a need for an advisory standing committee with a considerably wider remit than that of any body now in existence. This remit should include a responsibility for examining broad questions relating to intrinsic objections to current and future developments in the use of animals. We note, however, that many developments in this field raise questions across a whole range of topics. To take one example, the attempt to generate a line of transgenic pigs which could provide organs and tissues for transplantation into humans raises questions not only having to do with treatment of animals, but also questions about the acceptability of animal organs to the intended human recipients, the regulation of experimental treatments and the use of animals modified in such programmes as sources of food. There would be an advantage, therefore, in a standing committee being given responsibility to advise Ministers not simply in relation to animals, but in relation to the whole field of bioethics and the various questions which arise within it, questions which must be answered in the process of developing public policy. Since it is the Government which finally has responsibility for policy in these matters, it is for Government to establish a standing committee of the sort we envisage.

3.32 It is, of course, important that such a committee should secure the widest measure of public support. To that end it is essential that in its composition the committee should not only reflect the plurality of moral outlooks in our society, but that it should also be a forum for the rigorous development and examination of conflicting viewpoints.

3.33 A standing committee of this sort would be of assistance, we believe, to Ministers, but would also be welcomed by those engaged in scientific research and its practical or commercial application. The present lack of an adequate forum for the full and careful discussion of the complex questions raised by new technologies leaves a vacuum in which public concerns, such as those expressed in reply to our consultation letter, can go unaddressed. Public acceptance of legitimate applications of scientific research is severely threatened by a state of affairs in which ethical issues are in danger of being ignored rather than debated, and a forum for discussion of such issues has an important part to play in the creation of a greater trust between scientists, industry and the public.

3.34 In the light of these considerations we recommend the creation of a standing committee with the role and responsibilities we have outlined.

---

[5] An independent statutory committee established under ASPA to advise the Secretary of State on that Act.

CHAPTER 4:   BREEDING TECHNIQUES AND ANIMAL WELFARE

4.1   In this chapter we consider the implications for animal welfare of the use of advanced breeding techniques. This is an issue of considerable concern as responses to our consultation letter and general public discussion of these matters both show.

4.2   We have approached each technique in the following way. In the first place we have tried to understand as fully as is necessary what it involves for the animals concerned. Next we have considered the regulations which currently relate to the technique. The third task has been to establish the welfare problems, if any, to which the technique is thought to give rise. Then, in the fourth place, we have considered the adequacy with which the regulations meet the alleged difficulties, and have made recommendations in the light of this consideration.

4.3   In making these comments and recommendations we have recalled the principles set out in Chapter 2, and so have kept the following questions in mind: does the technique cause harm to the animals involved and, if so, is the harm of such a kind that it ought not to be inflicted on an animal? If it does cause harm, but is not so severe or of such a kind as to be straightforwardly impermissible, is the harm caused justified by a sufficiently substantial good? Further, if the harm is so justified, is it reduced, as far as is reasonably possible, by good practice?

4.4   Before turning to the individual techniques, however, it is necessary to mention the general welfare regulations relating to the rearing and keeping of farm animals which apply to all the techniques which we shall look at separately. A consideration of these regulations will also serve to set out the variety of means which Ministers have at their disposal to ensure the welfare of farm animals.

Welfare controls in general

4.5   The Agriculture (Miscellaneous Provisions) Act 1968 makes it an offence for any person to cause unnecessary pain or distress to any livestock situated on agricultural land. It is for the courts to determine in any case where an offence has allegedly been committed whether in fact unnecessary pain or distress was caused – this determination will be made by reference to evidence as to customary procedures and usual standards of competence.

4.6   Subject to Parliamentary approval, Ministers are empowered under the 1968 Act to make mandatory regulations in relation to animal welfare and also to issue codes of recommendations for the welfare of livestock. In exercise of the power to make mandatory regulations, Ministers have issued a variety of measures. These include those necessary to implement EC Directives on the welfare of battery hens, calves and pigs and the Council of Europe Convention (as revised) on the protection of animals kept for farming purposes which lays down general principles for the welfare of farm animals.

4.7   In exercise of the power to issue codes of recommendations Ministers have to date produced nine codes, covering cattle, sheep, goats, pigs, domestic fowl, turkeys, ducks, farmed deer and rabbits. The purpose of the codes is to encourage the highest standards of husbandry and in doing so they take account of five basic animal needs: freedom from thirst, hunger and malnutrition; appropriate comfort and shelter; the prevention, or rapid diagnosis and treatment of, injury, disease or infestation; freedom from fear; and freedom to display most normal patterns of behaviour. The codes not only provide authoritative guidance to farmers and others on how to ensure the welfare of their animals, but also have the backing of law in the following sense: although a breach of a provision of a code is not an offence in itself, such a breach can be used in evidence as tending to establish that the offence of causing unnecessary pain or distress to an animal under the Agriculture (Miscellaneous Provisions) Act 1968 has been committed. Presently only two of the codes (those concerning cattle and farmed deer) make specific recommendations on breeding, but it is obviously possible that such recommendations could appear in revisions of these codes, where appropriate.

4.8 In issuing codes or mandatory regulations, Ministers must consult with interested parties and seek Parliamentary approval. In such consultations, and on other matters relating to animal welfare, Ministers are advised in particular by FAWC, which was established in 1979 as an independent body with a membership which includes farmers, academics, veterinarians and those specifically concerned with welfare. Its role is to keep under review the welfare of farm animals on agricultural land, at market, in transit and at places of slaughter, and to advise Ministers on any legislative or other changes considered necessary. The Council has the authority to investigate any topic within its remit, to consult whomever it wishes, and to publish reports in whatever form it deems appropriate. Its advice to Ministers is usually in the form of published reports upon which Ministers consult before issuing a formal response. It has been the practice of the Government to take FAWC recommendations fully into account when determining future policy on animal welfare.

4.9 In addition to these general powers in relation to welfare, Ministers have power under Section 10 of the Animal Health and Welfare Act 1984 to make regulations controlling the artificial breeding of livestock (specifically the use of AI and the transfer of ova and embryos). The Bovine Embryo Collection and Transfer Regulations 1993 were made under powers granted by this Act. Regulations made under this Act also implement EU legislation concerning the animal health requirements for trade in, and import of, livestock semen and embryos.

4.10 Apart from offences created under the measures already mentioned, the various Protection of Animals Acts 1911–1988 make it an offence for any person to cause or fail to relieve the unnecessary suffering of any domestic or captive animal.

4.11 It is important to note, too, the Veterinary Surgeons Act 1966, which has the effect of protecting animals from unqualified practitioners. Acts of veterinary surgery are defined in the Act, and it is an offence for anyone but a veterinarian to carry out acts so defined unless specific exemptions have been made in accordance with the Act. As the statutory regulatory body for the veterinary profession, the RCVS can issue codes governing professional conduct and is required to investigate all complaints. Breach of a code is not a prerequisite for action by the RCVS, but breach of such codes would usually be a matter of professional misconduct and the subject of disciplinary actions. Where certain acts can only be performed by a veterinarian, the RCVS thus has the ability to establish principles of good conduct which should apply across the profession. It has to be remembered, however, that these codes are essentially advisory, and that the sanctions available to a disciplinary committee, such as having a veterinarian struck off, will be used only in grave cases. It cannot be assumed, then, that a professional code of practice will be as satisfactory a safeguard of animal welfare as mandatory regulations issued under an appropriate Act, though even where mandatory regulations exist, a professional code of practice can itself still serve a useful purpose in providing a supplement and support to the regulations.

4.12 It can be seen, then, that the welfare of farm animals is protected in three different ways: by broad legislative provisions making it an offence to cause unnecessary suffering; by more detailed regulations made under enabling Acts and relating to specific problems; and by the issuing of various codes, either statutory codes by Ministers or advisory codes by the veterinary profession, which encourage good practice in relation to animal husbandry or veterinary treatment. In making its recommendations, the Committee has taken note of the fact that in a field which is changing rapidly, the issuing of codes has the advantage of being a simpler and more flexible means of controlling developments than is the issuing of mandatory regulations. It also has the advantage over leaving things to the operation of the broader legislative provisions in that the issuing of codes is a way of actively encouraging good practices, rather than simply a means of punishing bad ones after the event. Having said that, however, there may be cases in which mandatory regulations are a vital means of protecting animal welfare.

The techniques

Selective breeding

4.13 Strictly speaking selective breeding falls outside the remit of this committee since we have been asked to consider ethical issues raised by emerging technologies in the breeding of farm animals – selective breeding has been used from earliest times when animals were first domesticated, so is not emerging, and in any case is not a technology. It is necessary, however, to make some remarks about selective breeding for the simple reason that the acceptability of the new technologies is in part related to the character and potential of selective breeding. In this regard two main points need to be made: in the first place, the development of both the science of genetics and sophisticated means of analysing data has rendered selective breeding a far more powerful means of improving farm animals than it previously was; in the second place, selective breeding is not invariably neutral as regards animal welfare.

4.14 For many centuries selective breeding was based on the simple notion of like begetting like. The limitations of this approach are obvious, for though some attributes – such as coat colour or presence or absence of horns – may be simply inherited and controlled by a single gene, many of the more important commercial characteristics such as growth rate or milk production are controlled by many genes, the expression of which may also be significantly influenced by the animal's environment. In addition, the identification of desirable animals from which to breed was somewhat rudimentary: the recognition of animals showing the preferred characteristics was dependent on human observation and memory rather than on scientific measurement, recording and analysis.

4.15 In the present century the development of the science of genetics has given an improved understanding of the mechanisms of inheritance. Alongside this improved understanding, the advent of computers and advances in statistical theory have made possible the highly sophisticated analysis of the performance records of animals and their progeny. Techniques in quantitative genetics such as best linear unbiased prediction (BLUP) and restricted maximum likelihood (REML) allow the assessment of many more individuals than was previously possible. Thus, for example, the performance of a bull can be assessed from the milk yield not only of the cows sired by the bull but also by reference to the milk yield of the daughters of these cows, their maternal sisters (i.e., sired by other bulls from the same mothers) and indeed all other female relatives no matter how distant. The techniques are particularly useful in relatively slower breeding species such as cows and sheep. Taken together, the ability to identify the most promising breeding stock by rigorous assessment and the greater insight into the mechanism of inheritance have made selective breeding far more effective in improving animals in respect of commercially important characteristics – though the gains in this regard are very significantly increased when this understanding and assessment is used in association with the techniques we shall shortly consider. Only in the case of pigs and poultry, producing as they do a high number of progeny each year, can rapid progress be made in genetic selection of commercially desirable strains of stock without resort to other techniques.

4.16 It is not to be supposed, however, that selective breeding, with or without the benefit of the developments we have considered, is without problems from the point of view of animal welfare. First, a breeder will have a limited range of animals from which to breed – the expense and difficulty of keeping some breeds of bull, for example, means that one or two animals may have to be used with a wide range of cows. This not only limits the effectiveness of selective breeding, but may lead to inbreeding or to other difficulties, such as problems with calving – because a bull which sires large calves was mated with a small cow, for example. Secondly, the intensive use of selective breeding to enhance a particular trait may result in unintended side-effects: leg weakness in broiler chickens, leg problems in sows and lameness and mastitis in dairy cattle are thought to have been caused in this way – though poor management may be a very significant factor in the problems found in dairy cattle. Similarly, some turkeys have been bred to such a size that they are incapable of mating naturally.

4.17 Selective breeding, then, can have highly objectionable side effects as regards animal welfare and it is important that FAWC exists to monitor these developments. The point here, however, is that it ought not to be assumed that the choice between selective breeding and the new techniques is a choice between a traditional method which is unproblematic, and newer methods which are fraught with problems from a welfare point of view. Indeed, it is worth noting here that some of the newer techniques may have an important part to play in preventing or undoing problems caused by selective breeding – thus, to take two examples, the wider choice of sire which AI makes possible can serve to prevent calving difficulties, just as some of the new techniques to which we now turn may contribute to the more speedy eradication of leg problems in broiler chickens.

The new breeding techniques

4.18 In cattle and sheep the usual rate of reproduction is one or perhaps two offspring per year (occasionally three or four in a few specialized breeds of sheep) so that the assessment of genetic merit was constrained even for males by limited information on progeny before the male was past its breeding prime. These limitations on rate of reproduction, in particular in cattle and sheep, were seen as major constraints on the improvement of these species for commercial use, and research addressed a number of techniques which could overcome them. The purpose was to facilitate more critical application of selective breeding rather than to attempt direct manipulation of genetic material. Some of these techniques may be utilized in application of genetically modified material but are not confined to that application and need to be recognized as distinct procedures.

Artificial insemination

4.19 AI was developed in the 1930s as a technique to control venereal disease and to facilitate selective breeding. Commercial application started in 1946 and became a successful world-wide business after it was found possible to preserve cattle semen in a deep frozen state. The technique is used extensively in the dairy industry where 67% of herds (by cow numbers) use only AI and a further 23% use some AI. Later, the technique was developed for pigs to maximize the benefit of males selected in sophisticated tests for genetic merit. It is also regularly used in poultry, in particular in turkeys, where there is virtually 100% use of AI. The technique is less usually applied to sheep because of their relatively low value, the physical difficulty of insemination, the extensive husbandry, and because of the marked seasonality in breeding which they display.

Artificial insemination in cattle

Technique

4.20   Semen for AI is collected by allowing a bull to ejaculate into an artificial vagina. This method is routinely used for bulls at AI centres and sometimes on farms. All bull semen is stored in a frozen state. Insemination is a skilled technique requiring training of the operator. The operator guides a catheter into and through the cervix of the cow (or heifer) using his other hand in the animal's rectum to help locate the cervix and to help direct the catheter. The procedure is relatively quick and anaesthetic or sedation is not usually required.

Current regulations

4.21   AI in cattle is controlled by regulations made under Section 10 of the Animal Health and Welfare Act 1984, which allows Ministers to make regulations controlling the artificial breeding of livestock. For cattle, these are the Artificial Insemination of Cattle (Animal Health) (England and Wales) Regulations 1985 as amended – similar regulations apply in Scotland and Northern Ireland. These regulations are due to be consolidated in the near future.

4.22   The main purpose of these regulations is to safeguard animal health, and to this end they provide for the licensing of premises where semen is collected, processed and stored. In order to obtain a licence, premises must first be approved by the competent authority. It is a requirement that to hold a licence to process semen a centre should be for the practice of AI only, be under the control of a veterinary surgeon on a day-to-day basis, have adequate livestock housing, have available a processing laboratory and that its staff and operation techniques meet required standards. AI centres are only able to accept animals which have been approved, and donors must undergo a series of health tests, set down in EC Directives and implemented in UK regulations, before they can be admitted.

4.23   In order to practise AI in cattle, non-veterinarians are required to be competent in the technique. Those that are employed in a commercial operation must be full-time employees of the holder of a supply licence, be competent in the technique and the associated hygiene precautions, and under the general direction of the veterinary surgeons specified in the licence. Persons carrying out AI solely for their own cattle must either have carried out AI regularly prior to the introduction of the 1977 Artificial Insemination Regulations, or have completed a course of training, under the direction of a veterinary surgeon and which was recognized by the former Agricultural Training Board. MAFF are currently reviewing the approval procedure to take account of changes in the structure of the industry, with a view to determining the appropriate bodies to grant certification in the future.

Welfare concerns

4.24   Concerns were expressed by some respondents about the implications of AI for the welfare of the animals involved. As regards bulls, concerns related to the conditions under which they are kept in AI centres as well as to frustration caused to the animal by the procedures by which semen is obtained – a bull is first 'aroused' by the presence of another animal, in order to improve the quality of the ejaculate, before being allowed to ejaculate into the artificial vagina. With regard to the insemination of cows, there was concern about the degree of discomfort resulting from the insemination itself, as well as from the restraint which may be necessary. This concern is augmented by the fact that its main practitioners are not veterinarians, but trained technicians. Some respondents thought that this situation could compromise the welfare of the animals involved.

Comments and recommendations

4.25 AI in cattle is a long-established procedure which is relatively straightforward. Provided that an animal is properly housed and handled it does not seem to us that the method by which semen is obtained, or its quality improved, threatens, or threatens to any significant degree, the welfare of bulls. We believe it is important, however, that the system of licensing be maintained. Although the licensing system is not intended primarily as a welfare measure, the requirements to be met in gaining a licence (specifically the requirement that an AI centre be under the control of a veterinary surgeon on a day-to-day basis) and the fact that such centres are subject to regular inspection by the State Veterinary Service, represent important safeguards of the welfare of the animals involved.

4.26 Similarly the discomfort suffered by cows being restrained and artificially inseminated is, we believe, likely to be at worst very slight, supposing that is that the procedure is carried out by a properly trained operative. It should be pointed out of course, that a cow may need to be restrained for natural mating, and that where a bull is oversized or multiple matings take place, discomfort is likely to result. It is nonetheless important that the highest standards of animal welfare are maintained wherever possible, and we recommend that in the review of the approval procedure there should be no relaxation of the requirement that non-veterinarians seeking to carry out this work should have reached a high level of competence.

Artificial insemination in sheep (and goats)

Technique
4.27 The collection and storage of semen in sheep closely resembles the practice in cattle – semen is collected from rams using an artificial vagina and can be frozen. Insemination, however, presents many more problems in sheep than in cattle because of anatomical differences. Although non-surgical intra-vaginal insemination can be used for ewes, it is not widely practised. Another non-surgical technique which is being explored is transcervical insemination, whereby semen is deposited into the uterus by means of a pipette inserted through the cervix. **Laparoscopic** (keyhole surgery) is the more usual technique, where semen is injected directly into the uterus by a pipette. The technique involves the animal being inverted in a restraining cradle, and administration of either general anaesthetic, or local anaesthesia usually with sedation. Small incisions are then made in the abdominal wall for the insertion of the laparoscope and the inseminating pipette. The technique for goats is similar.

Current regulations
4.28 Laparoscopy in sheep and goats is an act of veterinary surgery which, under the Veterinary Surgeons Act 1966, can be only be performed by a veterinarian. Though there are regulations relating to the health and trade aspects of AI in sheep and goats, which serve to protect these animals against disease, there are no further regulations relating specifically to the use of these techniques.

Welfare concerns
4.29 Concerns about the obtaining of semen would be essentially the same as those expressed in relation to bulls. As regards insemination, however, the issues are obviously different from those arising in the case of cattle. In the case of non-surgical AI, the handling of the animals can present problems, and with intra-vaginal AI the technique may bring risks of infection. The transcervical technique is also problematic: the insertion of a pipette through the cervix is very difficult to accomplish without causing damage (because of the complex, folded nature of the ewe's cervix) and laparoscopic AI, which may be less traumatic for the animal, is nonetheless an invasive procedure. Prior to surgery it may be necessary to withdraw food and water; during surgery the animal must be inverted and this may cause distress; the timing of the

administration of anaesthetic is critical; and as with any surgery, there is a risk of infection and other complications.

Comments and recommendations

4.30 The industry and the veterinary profession have rightly taken the view that the further development of non-surgical techniques is vital if AI is to have a wider role in the breeding of sheep and goats. In the meantime, there is a need for adequate control of these procedures to ensure that the welfare of the animals involved is not adversely affected.

4.31 In the case of the two non-surgical techniques, we note in the first place that FAWC considered transcervical AI in its 1994 report on the welfare of sheep[6] and recommended that this method should be used only by a qualified veterinary surgeon trained in the technique. We share their view that carried out inexpertly, transcervical AI represents a significant threat to an animal's welfare, and endorse their recommendation – our proposals in para. 4.33 seek to address this problem. Intravaginal AI, although not requiring the same degree of skill, nonetheless ought to be carried out properly and we recommend that those non-veterinarians carrying out this procedure should be required, as with cattle, to hold a certificate of competence obtained from an appropriate body (para. 4.24).

4.32 In relation to the surgical methods, a code of practice has been draw up by the Sheep Veterinary Society which recommends that laparoscopy and **laparotomy**, as invasions of the body cavity, should continue to be performed only by veterinary surgeons, who are competent in the procedure. The code also makes recommendations as regards the proper use of analgesia and the reducing of stress by good handling, and also suggests that the health, maturity and general suitability of the animal to be inseminated be carefully reviewed to enable a normal pregnancy and birth. Although we have not examined this code of practice in its details, it is apparent that it is the result of a careful and thorough examination of the subject and represents a basis for the establishment of proper welfare safeguards in relation to the practice of laparoscopic AI in sheep and goats.

4.33 Even where they are carried out according to the highest standards, however, both transcervical and laparoscopic AI in sheep and goats represent significant threats to an animal's welfare. We take the view that non-therapeutic surgery on animals requires special justification, and we are not persuaded that such procedures have a place in routine breeding programmes. In reaching this conclusion we have taken particular note of the concerns of the Sheep Veterinary Society. We therefore recommend that non-intravaginal AI should continue to be performed only by veterinary surgeons competent in the procedures and that, in consultation with FAWC and the RCVS in particular, the following provisions be introduced, by code of practice or where appropriate, by regulations under the Animal Health and Welfare Act 1984, to regulate non-intravaginal AI of sheep and goats, unless carried out under the ASPA:

(a) that laparoscopic AI be performed with appropriate and adequate analgesia;

(b) that in all cases of non-intravaginal insemination, a veterinary surgeon should review the health, maturity and general suitability of the animal to be inseminated in order to ensure, as far as possible, a normal pregnancy and delivery;

(c) that non-intravaginal AI be used only in disease control programmes and in recognised breed improvement schemes[7].

---

[6] Farm Animal Welfare Council, *Report on the Welfare of Sheep*, MAFF, 1994, PB 1755.
[7] This refers to such schemes as (a) those operated by sheep breeding organisations which have been approved under EC Decision 90/254, laying down criteria for approval of breeding organisations and associations which export or maintain flock books for pure bred sheep and goats (similar schemes exist for cattle and pigs); or (b) sire reference schemes using MLC recording schemes.

4.34 It will be clear that we have not attempted to formulate the details of the regulations we recommend – we have not, for example, considered what system of monitoring or reporting is necessary to ensure compliance with the regulations we envisage. These are matters on which Ministers will need to seek further guidance. We have, however, set out to the two broad principles which we regard as crucial in devising regulations in relation to surgical AI – that they should ensure both good practice in performance, and that it is used only where there is good cause.

4.35 It has been put to us that relative to the use of natural insemination, the number of pregnancies produced by AI is small, and that the economics of the keeping of sheep makes it likely that this will remain the case. Whilst we acknowledge this point, it does not constitute a reason against the making of regulations along the lines we have suggested in relation to both surgical and non-surgical techniques. In the first place the economics of the situation depends on many factors and may change quite quickly and in such a way as to render the use of these techniques financially advantageous. In the second place, though the number of ewes subject to AI is small relative to the total flock, this does not excuse a failure to ensure their welfare.

Artificial insemination in pigs

Technique
4.36 Semen is collected from a boar by allowing it to mount a dummy sow, probably after a false mount. After it has mounted, the boar's penis must be firmly held prior to and during ejaculation. Trained boars will donate semen readily and it is usually diluted and used within 2–4 days. While boar semen can be successfully frozen, the fertility of frozen semen, unlike cattle and sheep semen, is substantially lower than that of fresh semen. The restricted storage life of pig semen places a limitation on the general applicability of this technique.

4.37 Insemination is carried out by passing a spiral tipped catheter into the cervix and then allowing the semen to flow into the uterus. The procedure takes about 5–10 minutes, and a sow or gilt in **oestrus** will usually stand for the requisite length of time. Anaesthetic or sedation is not required, though consistently successful inseminations (i.e., leading to high conception rates) require skill and training. There are limitations on the commercial development of the technique due to, among other reasons, the difficulty of detecting oestrus in sows.

Current regulations
4.38 AI in pigs is controlled by the Artificial Insemination of Pigs (England and Wales) Regulations 1964 and the Artificial Insemination of Pigs (EEC) Regulations 1992. These regulations are under review and are due to be consolidated into one regulation in the near future. Separate but equivalent regulations apply in Scotland and Northern Ireland, and are also being reviewed. Though the main purpose of these regulations is disease control, they also require that AI centres for pigs are licensed on similar terms to AI centres for cattle (para. 4.23). This means that such centres must be under the day-to-day supervision of a veterinarian, and are subject to regular inspection by the State Veterinary Service. Since neither the obtaining of semen nor insemination itself are acts of veterinary surgery, both procedures may be carried out by non-veterinarians. Nonetheless the general welfare regulations referred to in para. 4.6 apply.

Welfare concerns
4.39 The welfare concerns as regards the collection of semen and insemination are essentially the same as those relating to cattle.

Comments and recommendations
4.40 Given that the concerns are essentially the same as those relating to cattle, it seems to the Committee that the same safeguards should be put in place. The use of AI

in pigs is not, and is perhaps unlikely to become, as widespread as it is in cattle. It does, however, have a place in specialized breeding programmes, and in circumstances where it is essential to reduce the risk of the spread of disease involved in the movement of live animals, or as a means of providing supplementary insemination at peak periods when boars would otherwise be over-used. Nonetheless it is appropriate and necessary that the practice, even if presently on a relatively small scale, should be regulated in such a way as to ensure as far as possible the welfare of animals involved. We therefore recommend to Ministers that the system of licensing AI centres should continue and that, as with cattle, non-veterinarian practitioners of the technique be required to hold an approval certificate to be obtained from an appropriate body (para. 4.24). Regulations to this effect could be made under the exercise of powers granted by the Animal Health and Welfare Act, 1984.

Artificial insemination in poultry

Technique
4.41 AI is mainly used for turkeys although it is used for chickens as well. For turkeys, semen is obtained from the male by pressing on the abdomen near the vent to 'milk' the semen. The semen is collected and used to inseminate the turkey hen by inserting a pipette into the cloaca and releasing the semen. The procedure in chickens is very similar.

Current regulations
4.42 There are no specific welfare provisions relating to artificial insemination of poultry, but the general welfare provisions will apply.

Welfare concerns
4.43 Artificial insemination in poultry requires the handling and restraining of both male and female birds and this is thought by some to threaten the animals' welfare. Concern may also relate to the collection of semen (which may occasionally cause the formation of small haemorrhages on a bird's abdomen) and to insemination itself.

Comments
4.44 Both the obtaining of semen and insemination are relatively simple procedures and pose no serious threat to the welfare of the animals involved, supposing that is that the animal is handled by a skilled operative in an appropriate manner and is accustomed to such handling. The formation of haemorrhages on the abdomen of the male bird is caused by the breaking of minor capillaries and should usually not be a matter for concern. It should also be pointed out that natural mating in turkeys may pose a welfare problem, since the stag, even in unimproved strains, claws the hen when mounting. Furthermore, the use of AI seems to reduce the risk of infection.

4.45 It may, however, be objectionable that conventional breeding has led to the production of turkeys of such a size as to be incapable of natural breeding. An objection may come from a concern for animal welfare – even if the turkeys experience no particular stresses in being prevented from breeding because of their size (and in any case they would be unlikely to be allowed to breed even if they were capable of so doing) it is hard to believe that the physiological transformation which has produced this incapacity is without other side effects which are significant for the animal's welfare. We are aware that FAWC is examining this matter and welcome the attention they are giving to it. But whether or not there is an identifiable welfare problem, the breeding of birds who are physically incapable of engaging in behaviour which is natural to them is fundamentally objectionable.

Superovulation/synchronization of oestrus

Introduction

4.46 The number of eggs available for fertilization can be increased by the administration of hormones in a technique known as superovulation. This technique can be used in association with embryo transfer (see next section), but can be used on its own to produce a larger number of offspring. Lower doses of hormone can also be administered, again not necessarily in association with other techniques, simply as a means of synchronizing oestrus and so of better managing reproduction. In sheep, for example, superovulation is used to increase twinning rate, and induced ovulation or oestrus synchronization is employed to ensure effective use of the ram and to shorten the lambing period. In cattle, induced ovulation is used to bring a cow into oestrus or to increase the likelihood of pregnancy following first service or AI. In pigs, synchronization of oestrus is also used as a means of managing reproduction more easily and of ensuring greater efficiency.

Technique

4.47 In sheep and cattle synchronization of oestrus can be brought about by an intra-vaginal progesterone device, a subcutaneous implant or a long acting injection. Alternatively, or in addition, prostaglandin injections may be given. In both cases, a further hormone injection is then given to stimulate ovulation and oestrus. In pigs the synchronizing drug is normally administered orally.

Current regulations

4.48 Superovulatory and oestrus synchronizing drugs fall within the terms of the Medicines Act 1968 and can be prescribed only by a veterinarian for animals under his care.

Welfare concerns

4.49 There may be some discomfort to the animal from the use of intra-vaginal devices and, particularly in sheep, from distress caused by their handling. There is a further worry that the use of superovulatory drugs may cause side effects, such as enlarged ovaries, and this may cause at least discomfort. There is also concern about the use of superovulation to produce multiple births. Although in lowland flocks and certain breeds of sheep there are generally few problems, and twinning may be both desirable from the farmer's point of view and unproblematic as regards the animal's welfare, this may not be the case in hill flocks or in cattle. Twin lambs or triplets, being smaller and more frail, are less likely to survive in upland conditions, and in cattle twinning can lead to problems at delivery, such as a longer labour or a retained placenta, and can result in lower milk production. Where twin calves are of different sexes, the female calf will usually be infertile.

Comments and recommendations

4.50 As a management technique superovulation and oestrus synchronization, as well as offering advantages to the farmer, may have welfare advantages to animals – in sheep, for example, the handling of the animals which the techniques necessitate also provides an opportunity to check that there are no feet problems or fly strike. It is difficult to assess whether some admitted side effects of the use of the drugs, such as enlarged ovaries, pose welfare problems – certainly we have not received evidence to persuade us that there is a welfare problem with side effects from the use of superovulatory drugs. This may, however, be an issue which FAWC should consider. FAWC may also wish to look at the issue of multiple births to establish whether or not there is a problem with the use of these drugs in this regard, or whether good husbandry and the administration of appropriate dosages are sufficient to ensure that there is no significant threat to an animal's welfare.

Embryo transfer

Introduction

4.51 Embryo transfer involves the removal of embryos prior to implantation from donor animals and transfer to recipient (surrogate) animals. Early work on the technique began in the 1950s, but involved full-scale surgical intervention; it was not until non-surgical techniques were developed for cattle in the 1970s that the procedures were used commercially. Even now embryo transfer takes place on a relatively limited scale: figures from the International Embryo Transfer Society suggest that there may have been 350,000 transfers in cattle world-wide in 1992.

Embryo transfer in cattle

Technique
4.52 In cattle, the production of fertilized embryos involves preliminary hormone treatment by injections over 3–4 days in order to stimulate multiple ovulations (superovulation) followed by natural service or AI. Seven days later, any resulting embryos are removed by flushing them from the uterus of the donor cow. The technique of flushing has basic similarities to AI, but takes much longer and is technically much more skilled. It involves passing a catheter through the vagina and cervix and along the uterus and then flushing the uterine lumen with specially prepared media. Posterior **epidural** (or general) **anaesthetic** is used. Most dairy cows if properly anaesthetized appear unconcerned by the procedure, but cows from beef breeds may be more fractious as a consequence of being restrained.

4.53 The collected embryos are graded for suitability for use before transfer to recipient animals. The recipients of embryos are often oestrus synchronized by hormone injection and/or vaginal device in order to improve the survival of the transferred embryo. Embryo transfer is most usually accomplished by a non-surgical method very similar to AI, but a high degree of skill is required to ensure that the embryo is deposited in the correct place. As in embryo collection a posterior epidural (or general) anaesthetic is usual.

Current regulations
4.54 The Bovine Embryo Collection and Transfer Regulations 1993, made under the Animal Health and Welfare Act 1984, control the collection of *in vivo* fertilized embryos and the transfer of embryos into recipients. They provide for embryo collection and transfer to be carried out by a veterinary surgeon, or by members of an approved bovine embryo team which must be headed by a veterinarian. Non-veterinarian members of the team may carry out a transfer only if they are competent and have been trained by the team veterinarian; such a transfer is carried out under the responsibility of the team veterinarian. The regulations make it mandatory to use a general or epidural anaesthesia prior to all collection or transfer of embryos. They also require the examination by a veterinary surgeon of recipient animals prior to transfer, to determine that the recipient is suitable to receive the intended embryo and that there is no reason to believe that the animal would not be able to bring the calf to term and calve normally. The Veterinary Surgery (Epidural Anaesthesia) Order 1992 enables trained

lay technicians to administer epidural anaesthesia and lays down the conditions of their training and qualification. The Bovine Embryo Collection and Transfer Regulations 1993 are backed up by the existence of a Code of Practice for Embryo Transfer in Cattle issued by the RCVS. This provides further guidance to veterinarians on ensuring the welfare of animals undergoing these procedures. As already mentioned, superovulatory drugs fall within the terms of the Medicines Act 1968, and can be prescribed only by a veterinarian for animals under his care.

Welfare concerns
4.55 Embryo transfer raises a number of welfare issues. In order to produce a number of embryos for collection, donors are usually given hormone treatment by a series of injections to bring about superovulation before insemination – as we have said, this can cause enlarged ovaries, though whether this is a matter of concern is difficult to assess. The cervix and vagina, and even the ovaries and fallopian tubes, can be damaged by inexperienced operators while passing the catheter in order to flush embryos. As far as transfer is concerned, the procedure takes place about a week after oestrus by which time the cervix has tended to become narrower making insertion of the catheter through the cervix more difficult. The use of a general or epidural anaesthetic, although now mandatory, also brings risks such as temporary paralysis of the hindquarters. Finally, the implantation of an unsuitable or inappropriate embryo into the surrogate can lead to calving difficulties.

Comments and recommendations
4.56 The welfare concerns mentioned above are, for the most part, addressed by the Bovine Embryo Collection and Transfer Regulations 1993, backed by the existence of the RCVS Code. The Code and the Regulations together ensure that an animal receives proper anaesthesia and that all those carrying out collection and transfer are properly trained. They also seek to ensure that surrogate animals are in good health and capable of bringing the embryo to term without difficulties. Thus the Code and Regulations represent a significant step forward in ensuring the welfare of the animals involved in embryo transfer. One point that is not covered is the administration of the hormones required to bring about superovulation, and as already said, FAWC may wish to consider whether there is a need for regulations in this area.

Embryo transfer in sheep and goats

Technique
4.57 Embryo transfer is also used to a lesser extent for commercial reproduction of sheep and goats, but with these species surgical intervention is required to recover the embryos. As in cattle, hormonal injections and vaginal inserts followed by AI are required to produce multiple fertilized embryos, which are then collected either using general anaesthesia and surgery (laparotomy), or the laparoscope method used for sheep AI described above, requiring general anaesthetic or local anaesthetic and sedation. Recipients are usually oestrus synchronized and transfer is by the same method as collection.

Current regulations
4.58 As in the case of cattle, regulations have been made under the Animal Health and Welfare Act 1984. However, the Artificial Breeding of Sheep and Goats Regulations 1993, though they require embryo transfer teams to be licensed and to comply with health precautions in the collection of ova and embryos laid down in an EC Directive, contain no further welfare provisions and relate only to intra-EU trade (i.e., they don't apply within the UK). Since collection and transfer are both acts of veterinary surgery under the Veterinary Surgeons Act 1966, they can only be performed by a veterinarian.

Welfare concerns
4.59 The practice of embryo collection and transfer in sheep raises many of the problems we have mentioned in connection with laparoscopic AI: both the collection

and transfer of embryos constitute major surgical interventions with attendant problems. We note the concern of the Sheep Veterinary Society, and the Code of Practice which they have drawn up, to which we have already referred, and which relates to both laparoscopic AI and embryo transfer. The use of superovulatory drugs may or may not raise welfare questions, but the poor fertility of superovulated ewes means that they are likely to be impregnated by AI rather than natural service, thereby undergoing two invasive operations.

Comments and recommendations

4.60 The comments we made in relation to AI in sheep are appropriate in relation to embryo transfer too, and our recommendations are therefore parallel with those already made. Specifically we recommend that embryo transfer in sheep and goats should only be performed by veterinary surgeons competent in the procedure and that, in consultation with FAWC and the RCVS in particular, the following provisions be introduced, by code of practice or, where appropriate, by regulations under the Animal Health and Welfare Act 1984, to regulate embryo transfer in sheep and goats, unless carried out under the ASPA:

(a) that embryo transfer be performed with appropriate and adequate analgesia;

(b) that in all cases a veterinary surgeon should review the health, maturity and general suitability of the animal to receive an embryo, to ensure as far as possible a normal pregnancy and delivery;

(c) that embryo transfer be used only in disease control programmes and in recognised breed improvement schemes[8].

Embryo transfer in pigs

Technique

4.61 Embryo transfer in pigs is rare, because pigs are so easily bred using natural service or AI. The main reason for using it is to introduce disease free strains or new bloodlines in specific herds. The collection and transfer procedure is similar to that for sheep. However, general, rather than local anaesthesia, prior to surgery (laparotomy) is required to recover the embryos. Pigs can be successfully superovulated if necessary to produce a larger number of embryos than normal, but since pig embryos are difficult to freeze successfully, commercial embryo transfer is not widely used. Were this difficulty overcome, and were attempts to develop per vaginam methods of collection and implantation successful, embryo transfer in pigs may become far more common.

Current regulations

4.62 There are at present no specific regulations controlling embryo transfer in pigs, though whilst collection and transfer are surgical procedures, they may only be performed by a veterinarian. Embryo transfer regulations will be included, however, in the review of the AI in pigs regulations mentioned above in para. 4.37.

Welfare concerns

4.63 The surgical operations required to collect and transfer embryos are significant interventions with associated risks and the welfare concerns in relation to embryo transfer in pigs are essentially the same as those discussed in relation to sheep and goats.

---

[8] This refers to such schemes as (a) those operated by sheep breeding organisations which have been approved under EC Decision 90/254, laying down criteria for approval of breeding organisations and associations which export or maintain flock books for pure bred sheep and goats (similar schemes exist for cattle and pigs); or (b) sire reference schemes using MLC recording schemes.

Comments and recommendations

4.64 The technical difficulties involved in embryo transfer in pigs, as well as the ease of breeding by natural service, makes it highly unlikely that there will be any significant demand for the use of embryo transfer, in its present form, apart from in the context of research. We believe that it is important that laparotomy, as an invasion of the body cavity, should continue to be performed only by veterinary surgeons, and that good practice in relation to analgesia, selection of animals, and so on, should be observed – here too, then, the Code of Practice drawn up by the Sheep Veterinary Society provides a model of what is needed to ensure the welfare of animals subject to these procedures. We make the point again, however, that whether or not carried out in accordance with the highest standards, laparotomy, as a non-therapeutic surgical intervention, is not justified in routine breeding programmes. We therefore recommend that the regulations we have already outlined in para. 4.59, in relation to embryo transfer in sheep, apply also to pigs.

4.65 If, however, difficulties in freezing pig embryos were overcome and per vaginam methods of transfer of embryos were developed, commercial interest in the wider application of embryo transfer would probably be strong. In the event of these developments, we recommend that, as with cattle, an approval procedure should be established to ensure that technicians seeking to carry out this work should do so under veterinary supervision and should have reached a high level of competence before they receive a licence.

Embryo transfer in other animals

4.66 It is possible to use surgical embryo collection and transfer techniques in deer. The nervousness of deer make them highly unsuitable subjects for any such techniques, which raise, in addition, all the welfare issues already discussed in relation to sheep, goats and pigs. We take the view then, that there is no place for these techniques in routine breeding programmes, and recommend accordingly that the regulations we have outlined in relation to these other species be extended to protect deer.

4.67 It is not possible to superovulate horses to produce multiple embryos, although in some countries embryo transfer is used with some breeds of horses to allow the non-surgical collection and transfer of a single embryo. The procedure is very similar to that used for cattle. There are no regulations relating to AI or embryo transfer in horses, though the issue is being considered in the EC. Whether or not embryo transfer in horses is deemed to be an act of veterinary surgery within the terms of the Veterinary Surgeons Act 1966, the value of the animals involved makes it likely that these techniques will be performed by a veterinarian whose competence and training will ensure that the welfare of the animals involved is safeguarded. Further, the commercial pressures which might lead to poor practice in embryo transfer in cattle are unlikely to exist in relation to horses. Nonetheless, those seeking to carry out AI or embryo transfer in horses should, as much as those carrying out the work in cattle, be competent to do so, and we recommend that regulations be made under the Animal Health and Welfare Act 1984 requiring that non-veterinarians should carry out this work under veterinary supervision and only if they hold an approval certificate granted by the appropriate body (see para. 4.24) after suitable training.

Ultrasound scanning

Technique

4.68 A technique relevant to both AI and embryo transfer is the development of the use of ultrasound scanning for pregnancy confirmation. In cattle and mares the procedure requires a hand held probe to be inserted into the animal's rectum and manoeuvred in order to scan the uterus. For sheep and pigs the abdomen is scanned externally, in the same way as pregnant women are scanned.

Current regulations

4.69 At the moment there is no specific legislation although general welfare controls will apply.

Welfare concerns

4.70 The RCVS is concerned that in unskilled hands that there could be a threat to animal welfare through use of the rectal probe either because of the possibility of damage to surrounding tissues, or because of misdiagnosis and consequent mis-management. The RCVS has requested MAFF to consider regulating the procedure in bovines if it is to be performed by non-veterinarians.

Comments and recommendations

4.71 There is some controversy about whether there is a problem with this technique. We recommend that FAWC review the evidence on this issue, and if necessary regulations be made, as appropriate.

In vitro fertilization, semen and embryo sexing, and cloning

4.72 AI and embryo transfer are techniques which optimize selective breeding by allowing selection of both males and females as well as shortening the generation time. A number of techniques are also being developed to increase the number of embryos available for transfer.

In vitro fertilization

Techniques
4.73 *In vitro* fertilization (IVF) is one of the ways of producing large numbers of embryos and is currently used on a small scale in the UK and elsewhere, principally in relation to cattle. Immature oocytes are matured in a laboratory after collection from slaughtered cows at abattoirs. The oocytes are then fertilized using selected semen, cultured for a further period and then transferred to recipients. Oocytes for IVF can also be recovered from live cows (ovum pickup), by direct aspiration from the mature ovarian follicle by means of a needle passed through the wall of the vagina under ultrasound guidance. As it is a very skilled technique it is in limited use.

Current regulations
4.74 Ovum pickup is an act of veterinary surgery within the meaning of the Veterinary Surgeons Act 1966, and can only be performed by a veterinary surgeon. The transfer of any embryos produced by these means is governed by the Bovine Embryo Collection and Transfer Regulations, 1993.

Welfare concerns
4.75 Though ovum pickup is an act of surgery within the meaning of the Veterinary Surgeons Act, it is a relatively non-invasive intervention. Its use does, however, raise welfare issues and the fact that it may be performed as frequently as twice a week on a particular animal is cause for concern. Furthermore, the transfer of embryos produced by IVF has led in some cases to calves which have a higher birthweight than normal for

the breed, causing problems for the mother at calving. The cause of these difficulties is not established, but it may occur because the embryos have been cultured in a particular medium.

Comments and recommendations
4.76 Though ovum pickup is not highly invasive, it is a non-therapeutic surgical intervention and we believe that its place in routine breeding is questionable. We are concerned, in particular, that such a procedure, with its attendant risks, should be regularly performed on a particular animal. We have not been able to consider this issue in as much detail as it deserves. Therefore we recommend that FAWC should consider the issue of the use of this technique.

4.77 On the matter of IVF it seems doubtful that a veterinarian could properly implant an embryo produced by certain current IVF procedures given the requirement in section 10(3b) of the Bovine Embryo Collection and Transfer Regulations that he should only implant an embryo where he "knows of no reason . . . which would cause him to believe that the animal would not be able to carry to term a normal calf . . . and to calve naturally." That is to say, the fact that an embryo has been produced by IVF is in itself cause for believing that the embryo could not be carried to term without difficulties. Since some techniques of culturing embryos do not seem to cause calving difficulties, however, a veterinarian who was satisfied that the embryo had been cultured in such a way could appropriately carry out the transfer. Provided that veterinarians ascertain the method of culture used in IVF, the Bovine Embryo Collection and Transfer Regulations offer a sufficient safeguard of the welfare of surrogates.

4.78 If and when the problem with oversized calves has been overcome, IVF may be useful in providing more embryos for transfer than could be supplied by current methods of embryo collection. In so far as these current methods of embryo collection themselves pose welfare difficulties, particularly where the collection of embryos is a surgical technique, the use of IVF may benefit animal welfare. The welfare problems to do with the transfer of embryos will, however, remain.

Semen and embryo sexing

Technique
4.79 The determination of the sex of the progeny of their livestock would be of considerable value to certain farmers. Beef producers, for instance, would like to breed chiefly male animals as more suitable for their purposes, whereas dairy farmers require a proportion of females for herd replacements. Were it possible to ensure the production of animals of the right sex in the right proportions, there would be significant gains.

4.80 Sex determination has been approached in two different ways: by the sexing of embryos and by semen sexing. The sex of embryos produced by IVF or recovered from the uterus can be determined by detecting the presence of the male or 'Y' chromosome. Embryo sexing has been available commercially for several years but the costs of the procedure and the fact that about half of the embryos will be of the 'wrong' sex has meant that it is not widely used. Semen sexing can be carried out by using sophisticated computer controlled sorting equipment, which sorts semen into the 'Y'-only bearing sperm (male) and 'X' bearing sperm (female). The sorting technique exploits the fact that the 'X' bearing sperm is fractionally heavier than the 'Y' bearing sperm. This sorted semen can then be used in the IVF process to make embryos of pre-determined sex.

Current regulations
4.81 This practice is not governed by regulations.

4.82 There are no welfare issues, though the destruction of unwanted embryos in embryo sexing may be held to be objectionable since it involves large scale destruction of living organisms.

Comments
4.83 Current sorting procedures are not very effective at sorting sperm and produce such small volumes of sperm that they can only be used with IVF. An improvement in the sorting methods, which made them more effective and allowed sorted sperm to be used in AI, would render sperm sorting commercially more viable, and should be welcomed.

Cloning

Technique
4.84 Multiple genetically identical animals can be produced by cloning. This is done in four ways, as follows.

(a) When the embryo has reached the 8–16 cell stage it can be broken down (disaggregated) into individual cells, each one capable of developing into a full embryo, which can then be transferred to a recipient. This process is inefficient because not all the cells develop and therefore is not used commercially.

(b) A more successful cloning technique is to cut the embryo in half (embryo splitting) and transfer each half to a recipient. The result is identical twins.

(c) Another way of cloning, still at an experimental stage, is **nuclear transplantation**. Embryos of a slightly larger size (16–64 or more cells) are disaggregated to individual cells and the genetic material from each cell transferred to a mature egg which has had its own genetic material removed. The aim of this technique is to produce a large number of genetically identical embryos, but it is presently not very successful.

(d) Finally, certain cells known as embryonic stem cells can be removed from an embryo and multiplied *in vitro* to provide a source of nuclear material for the nuclear transplantation technique. This is still at an early experimental stage.

Current regulations
4.85 There are no specific regulations covering these techniques though the transfer of any embryos produced by these means is governed by the Bovine Embryo Collection and Transfer Regulations, 1993.

Welfare concerns
4.86 There have been problems with embryos produced by nuclear transplant or splitting causing overlarge calves and subsequent calving difficulties and the industry has stopped using these particular techniques for the moment. Concern has been expressed that the production of large numbers of animals which are genetically identical could have implications for the spread of genetic disorders, and for general susceptibility to disease.

Comments and recommendations
4.87 As with IVF, there may be gains to animal welfare if the use of these techniques obviates the need to resort to surgical collection of embryos. However, the problems with over-sized progeny are a serious cause for concern: in cattle the problem is

adequately addressed by the Bovine Embryo Collection and Transfer Regulations, 1993, but in sheep and pigs no such regulations obtain, and it is important that they should. We believe regulations we have recommended in paras. 4.59 and 4.63 would cover this point. As regards the alleged risks involved in the production of genetically identical stock, these seem to us to be illusory. Cloning might be used by elite breeders in order to make widely available a commercially superior animal, but genetic variation would be maintained in elite herds so as to allow for further genetic progress. Cloned animals will be prey to disease just like any other, but if they showed a particular susceptibility to disease, genetic or otherwise, they would no longer be used. It is in any case unlikely that in the foreseeable future cloning would be used on a sufficient scale to influence the health status of the whole animal population. These areas are dealt with further in Chapter 7.

Genetic modification

Technique

4.88 New scientific techniques now allow direct modification of the genetic material of animals in a way which does not occur naturally by mating (and/or natural recombination). These techniques permit the isolation and genetic modification of the DNA sequence of a gene in the laboratory (*in vitro*). Once modified the gene can be multiplied (or cloned) extensively in culture and purified. Genes can also be coupled to a variety of controlling sequences which regulate their expression in different tissues in response to different hormones or at different times in development. Such a 'hybrid' gene can therefore produce an altered protein with an altered action and regulated in a novel way. To produce transgenic animals, the novel gene has to be introduced into an animal's germline; genes from the same or a different species can be used. The classical, and still most often used, technique in laboratory and farm animals is to inject several hundred copies of the novel gene into the pronucleus of a single fertilized oocyte. The injected gene incorporates into the host's DNA randomly and as the embryo divides, it is copied into every cell including the germline and hence can be passed from generation to generation. Once injected the embryo is cultured *in vitro* for 24 hours or so and then implanted into a surrogate mother. The overall success of the method is 0.5–2.0% of the oocytes injected.

4.89 A further approach, which has been successful in mice and is being developed in farm animals, aims at overcoming the two drawbacks of the direct injection approach: the random incorporation of the injected gene and the low success rate. This involves growing embryonic stem cells in culture and performing the technique of gene transfer in the laboratory (*in vitro*). In principle this could increase the success rate to 100% and the novel gene could be targeted into a specific place in the host's DNA – usually the host's own homologous gene. Once fully developed, this approach would both dramatically reduce the cost of transgenesis and improve the specificity and range of alterations possible.

4.90 The remaining obstacle to using transgenesis to alter commercially important traits in farm animals, is our lack of knowledge of the genes involved. The traits are genetically complex and controlled by many genes; however co-ordinated programmes have been established within the EU to identify, clone and analyse trait genes and though they are still at an early stage, some trait genes have already been identified.

Current regulations

4.91 In the UK genetic modification of animals is a regulated procedure under section 2(3) of the ASPA, and may be undertaken only after the grant of a project licence from the Secretary of State[9]. In granting the licence the Secretary of State is required, amongst other things, to "weigh the likely adverse effects on the animal concerned against the benefit likely to accrue as a result of the programme" for which permission is sought. (In the terms we have used, this provision requires that an animal should be used in an experiment only if the good which is reasonably expected in carrying out the research is sufficiently substantial to justify the harm which the animal may suffer.)

4.92 Where a licence is granted and, subject to other legislation on genetic modification, research work goes ahead, section 15(1) of the Act requires that any protected animal which, at the end of the regulated procedures "is suffering or likely to suffer adverse effects", must be killed and so cannot be released from the control of the Secretary of State.

Welfare concerns

4.93 Whilst genetic modification may allow improvements in animal welfare (in relation to disease resistance, for example) or may be neutral in this respect, it may also, doubtless, be used (intentionally or unintentionally) in ways which harm animal welfare. There is, however, nothing in the techniques themselves which has negative welfare implications. Transgenesis may result in the infamously deformed Beltsville pigs[10], but it may equally produce the Edinburgh sheep (para. 3.21) which are seemingly unaffected by modification. In this respect these techniques are on a par with standard selective breeding which may or may not have deleterious effects on animal welfare.

4.94 Whether a proposed modification will be harmful is presently difficult to predict. The current lack of knowledge about animal genomes to which we referred above, contributes to the difficulty and means that any modification risks producing an animal whose welfare is in some way harmed. There are some 50,000 to 100,000 expressed genes in mammals and the effect of an inserted gene will depend on its position and on its interaction with other genes. Ideally the inserted gene will have only a limited effect in the specific area of metabolism or development in which it operates, and this seems to be the case with the Edinburgh sheep. However, in the case of the Beltsville pigs the modification was far from being neutral in respect of the animals' welfare. Copies of the human gene for growth hormone were inserted but were not subject to the normal physiological control, so that excess growth hormone was produced. The resulting animals were crippled by arthritis, were unable to reproduce and suffered other side-effects.

4.95 It is important to note that whilst in certain cases the harm which has been caused to an animal by a genetic modification may be so severe as to be apparent very quickly, in other cases it may only emerge as the animal is subjected to the varied conditions of life on a farm. Or it may be that the deleterious nature of the modification is apparent only in **homozygous**, not **heterozygous**, individuals.

4.96 Genetic modification may also affect the growth of an embryo such that it becomes a welfare problem to the surrogate mother. For example, if the modification causes unusually large calves, the recipient may not be able to give birth naturally although the calf itself is not adversely affected by the modification.

Comments and recommendations

4.97 It is ironic perhaps, that though genetic modification is the subject about which respondents have expressed the most disquiet, the welfare of transgenic animals seems

---

[9] The Home Secretary or the Northern Ireland Department of Health and Social Services.
[10] Transgenic pigs produced at the United States Department of Agriculture Agricultural Research Service in Beltsville, Maryland, USA.

to be more strictly safeguarded by current regulations than is the welfare of animals bred by other techniques. Since a proposed modification constitutes a regulated procedure under the ASPA, it is, as has been explained, subject to the test laid down in Section 5(4) before permission is granted for it to be attempted. That is to say, the Secretary of State must weigh the likely adverse effects on the animal against any likely benefit. Since, as we have said, the present ability of researchers to control the expression of a gene is limited, it is far from easy to be sure how likely adverse effects may be.

4.98 Some would wish to see a more prohibitive provision here – it might be said, and we would be inclined to agree, that where an experiment risks serious and profound harm to an animal that is reason enough to warrant its prohibition – as would be the case with experiments on humans. Even those who do not go along with such a strong principle might hold that if the harm which was done to the Beltsville pigs, for example, was not merely risked by a particular experiment but was a foreseen or reasonably likely consequence of it, such an experiment ought to be forbidden no matter any good results which are reasonably anticipated. But even if the ASPA included such a provision, it would not necessarily protect animals from serious harm; the difficulty is that the consequences of the Beltsville experiment could not have been foreseen or judged likely – the equivalent experiment in mice seems not to have had such dire effects on the animals.

4.99 Given the difficulty in anticipating adverse effects in experiments of this kind, it is important that the ASPA has a second element which is relevant to genetic modification, and that is section 15(1) of the Act which requires that any protected animal which, at the end of the regulated procedures "is suffering or likely to suffer adverse effects", must be killed and cannot be released from the Act's control. This means that even were the Secretary of State to grant a licence for an experiment which produced transgenic animals whose welfare was seriously impaired, the harm done to the modified animals would prevent their being released from the control of the Act and passing into commercial conditions. Indeed, once the side-effects of the modification became clear, the Inspector would be likely to recommend the immediate cessation of the experiment and the killing of any affected animals, or could require these steps under section 18(3) of the Act if he judged the animal to be undergoing "excessive suffering" – and in the case of the Beltsville pigs such a judgement would have been warranted.

4.100 It is important to note too, that section 2(3) of the Act makes anything done for the purpose of, or liable to result in, the birth or hatching of a protected animal a regulated procedure if, as a result, that animal is likely to be caused pain, suffering, distress or lasting harm. Thus the breeding of offspring from genetically modified animals is itself a regulated procedure, and the Home Office has made it clear that it will require the breeding of genetically modified animals to two generations of homozygosity before being satisfied that a modification can be regarded as one which does not cause "adverse effects".

4.101 If the phrase "adverse effects" is taken in a wide sense, then section 15(1) of the Act represents an important safeguard of the welfare of genetically modified animals and their progeny. It would mean, for example, that animals subject to the sort of modifications we held to be intrinsically objectionable in Chapter 3, as well as animals more obviously suffering because of genetic modification, could not be used for commercial purposes.

4.102 There are, however, at least two grounds for concern as to the adequacy of the protection that the ASPA affords. In the first place, the interpretation and application of the phrase "adverse effects" is a matter of some uncertainty. It may be that the modifications we considered objectionable (such as the one aimed at reducing the sentience of a pig so as to increase the efficiency of its conversion of food), would not be covered by the interpretation of this section. Or it may be that the range of welfare and other indications which are considered in the application of section 15(1) are insufficiently wide, and that those which are considered are not assessed with the scientific rigour which is in some cases possible and appropriate. Or it may be that animals are not observed in the range of conditions they may experience in commercial settings, but only in the highly controlled and protected settings of experimental research institutions. At root, then, the difficulty is that what is meant by "adverse effects" in section 15(1), and how the existence of such effects is determined, are matters which are within the discretion of the Secretary of State, and, in turn, those who administer the Act. This discretion is doubtless exercised after the widest consultation and consideration, but it remains the case that the adequacy of the Act in relation to the problems posed by genetic modification is not easily established.

4.103 In the second place, it may be thought that the ASPA only protects animals which are produced in UK laboratories. There is concern then, that there is nothing to stop the importation and use for agricultural purposes of an animal, such as the Beltsville pig, which would not have been released from the Act had it fallen under its terms.

4.104 The first concern could be dealt with were the Home Office to give an account of how the existence of "adverse effects" is established, and an indication of whether genetic modifications which could be judged intrinsically objectionable would be held to have caused "adverse effects". We recommend, therefore, that the Animal Procedures Committee[11] be invited to address this issue – our view is that if "adverse effects" are held to include damage to the natural integrity of the animal subject to modification, then the ASPA is sufficient to preclude intrinsically objectionable genetic modification.

4.105 As regards the second concern, any animal imported into the European Union or being moved about the Union must comply with the regulations concerning animal health and welfare which require certification of the animal's health status; as part of the certification procedure the veterinary inspector ensures the animal is fit to travel. If the animal is genetically modified and is being imported either for deliberate release or to be placed on the market, as defined, the appropriate prior consent from the authorities would be required (see Chapter 6 on risk). If the animal is being imported for contained use purposes, the regulatory controls are as described in Chapter 6. In either case the general welfare regulations will apply as appropriate. If the animal is imported for experimental purposes the terms of the ASPA will be relevant.

4.106 We note that as things stand, animals produced by transgenic modifications are more thoroughly protected than animals produced by conventional means. Conventional breeding, and the other techniques we are considering, could, intentionally or unintentionally, produce animals whose welfare is adversely effected. Unless the aim is to produce such animals for scientific use, programmes of breeding using such methods are not subject to the so-called 'cost-benefit' analysis which is required before the licensing of a proposed genetic modification, nor are the animals bred by these programmes protected by provisions such as are found in the ASPA.

Conclusions

4.107 In Chapters 2 and 3 of this report we set out the principles which properly govern the treatment of animals. We have maintained that these principles provide a

---

[11] This is an independent statutory committee established under the ASPA to advise the Secretary of State on the operation of that Act.

proper basis on which to evaluate the emerging technologies and their implications for animals, and our recommendations in this chapter, based on an application of these principles, fall into three classes.

- First, we have identified some uses of the new technology as intrinsically objectionable – in particular those instances of genetic modification which can be thought to constitute an attack on an animal's essential nature. We have had to consider whether the relevant current regulations (i.e. the ASPA) serve to prohibit such modifications. Our view is that these regulations are broadly speaking adequate, but that clarification is required on two points – the interpretation of the key notion of "adverse effects" and the means by which the existence of such effects is determined.

- Secondly, we have identified other uses of the new technology, which are not absolutely impermissible, as nonetheless justified only in particular circumstances where a substantial good is expected. We do not believe that non-therapeutic surgery is justified in routine breeding programmes, and therefore we have recommended regulations in relation to AI in sheep, goats and deer, and in relation to embryo transfer in those same species and in pigs, with a view to prohibiting the routine use of these techniques.

- Thirdly, we have judged certain uses of the new technologies to be generally acceptable, and have then considered whether current regulations ensure, as they should, that any harms caused in such cases are minimized. Some of our recommendations in relation to AI, embryo transfer, superovulation and induced oestrus, and ultrasound scanning, come under this heading.

CHAPTER 5: ADVANCED BREEDING TECHNIQUES AND THE USE OF PATENT LAW

5.1 The use of patent law in relation to advanced breeding techniques (specifically in relation to genetic modification) is the subject of much concern as the response to our consultation letter showed. In considering this concern it is necessary to understand something of the patent system and its rationale.

5.2 A patent can be seen as a form of contract between an inventor and society, in which certain limited monopoly rights are granted in return for the publication of information specifying the nature of the invention. In order to obtain a patent, under the European Patent Convention and the UK's Patent Act 1977, a process or product must be novel, non-obvious (i.e. there must be an inventive step) and capable of industrial application. Even if it satisfies these formal requirements, however, a product or process cannot be patented if it is likely to encourage offensive, immoral or anti-social behaviour, though this condition is interpreted very narrowly indeed – to such an extent that though many commentators agree that a letter-bomb could not be patented, no case can be cited where an application has been turned down by the UK Patent Office on moral grounds. If granted, a patent is valid for 20 years, and entitles the owner of the patent to prevent anyone else from exploiting the invention. It does not, however, entitle the owner of that patent to exploit it, and in particular confers no exemption from other legislation which may relate to the invention.

5.3 A patent system is said to have two principal merits. In the first place, it encourages research by ensuring that those who invest time, money and energy in devising novel processes or products can, subject to any other relevant regulations, obtain a return on their investment by exploiting their invention free of competition or by permitting others to exploit it under license in return for payment. In the second place, by requiring the disclosure of detailed information for the obtaining of patents, the system ensures that knowledge of technological advance is made widely available, both to other researchers and members of the public. Thus, were there no possibility of obtaining patents for inventions, research would be discouraged or would take place only in conditions of absolute secrecy. In either case, technological and scientific progress would be hampered. It is worth adding that in the case of research the safety of which may be a matter of concern, the degree of openness which patenting secures is a matter of some importance.

5.4 The patent system is not without detractors. Some point out that the very possibility of obtaining a patent discourages the sharing of information in the early stages of research, and thus may actually prevent the rapid advance which would occur where ideas were freely shared. Others allege that the advantages of having a product on the market first are sufficient to justify investment in research and development, and that technological progress would not be checked even though were there no patent system.

5.5 We are inclined to believe that the case in favour of a system of patenting is stronger than the case against, but the task of this Committee has not been to consider this wider issue, but a narrower question: whether there are specific reasons for excluding from the scope of this system the modifications of animals which the new biotechnology has rendered possible.

5.6 Though it is not possible under existing UK law to claim a patent on animal varieties or on offspring produced as a result of 'essentially biological processes' (for example, those produced by normal breeding), it is possible to obtain a patent for animals if an inventive step has been involved in obtaining it. Thus where an animal has had its genetic identity modified by artificial techniques, a patent may be claimed. The patent may relate to the modified gene or genes (the gene construct), the process by which the modification is made and/or the modified gene or genes are inserted into the target genome, and the resulting transgenic animal. The patent could be claimed for a single step in the process if it were inventive. The same patent could also be claimed (in subordinate claims) for that step in combination with other steps – up to and including the whole process. The scope of the patent will depend on how it has been drawn up and each patent claim is dealt with on a case-by-case basis.

5.7 Is this state of affairs satisfactory? Those who have replied on behalf of the biotechnology industry have naturally claimed that it is, and that any attempt to single out inventions in this area and deny them the patent protection which is given to other technologies should be resisted. It is argued that without such protection investment in biotechnology will be discouraged; that any work which did nonetheless go on in this area would necessarily be conducted in secret; and that were the UK or EU to limit the scope of patent protection so that work aimed at genetic modification were not capable of protection, then such work would simply move to countries with more favourable patenting regimes with resulting economic loss. It is further argued that any attempt to import moral considerations into the patent system (allowing grants of patents to some, but not all, biotechnological advances) should be resisted as adding an inappropriate element. The patent system is not the right forum, it is said, for achieving moral or social objectives such as the protection of animal welfare.

5.8 Those who are unhappy with the present situation express a number of different concerns. Some argue that extension of patenting rights to animals is offensive as such, in that it encourages us to view and treat animals as if they were simply products of human ingenuity or inventions, and thus to confuse living things with artefacts. Others, who may not object to patenting of animals on these grounds, but who are opposed to all experiments on animals, or who believe that genetic modification is in general likely to be harmful to animal welfare, contend that a denial of the possibility of obtaining patents will be an effective means of discouraging such work. And there is also the concern that patenting will allow the concentration of the ownership of genetically superior stock in fewer and fewer hands, to the detriment of Third World farmers and perhaps also to the detriment of smaller farming enterprises in general, and that for this reason it should be disallowed.

5.9 The same arguments which have been addressed to this Committee have also been advanced in the European debate which was initiated by the need to clarify and harmonize the scope of national patent laws in this area and which has led to an EU proposal which has been under consideration since 1988. The draft Directive on the Legal Protection of Biotechnological Inventions provides in Article 2.3 that "inventions shall be considered unpatentable where publication or exploitation would be contrary to public policy or morality" (a principle already enshrined in existing European and UK law), and clarifies this provision by stating that amongst such unpatentable inventions are "processes for modifying the genetic identity of animals which are likely to cause them suffering or physical handicaps without any substantial benefit to man or animals, and animals resulting from such processes."

5.10 It is clear, then, that the European Union Council of Ministers has declined to accept the arguments of either of the extreme positions in relation to this question. That is to say, it has been persuaded neither by the arguments which hold that biotechnological inventions should be patentable without further question if they meet the formal criteria (i.e., they are novel, non-obvious and capable of application), nor by the contrary arguments which would exclude all such inventions from patentability.

5.11 This Committee is in turn persuaded that this is the right approach. We can explain our view by considering the arguments which have been put to us, taking first of all the views of those who would exclude genetically modified animals from patent legislation altogether and reviewing alongside them the opposing positions.

5.12 As we have explained, a patent may relate to a modified gene, the process by which the modified gene is made and/or inserted, or the resultant transgenic animal itself. It is claims to patent transgenic animals which arouse most concern and it is contended by some that the patenting of genetically modified animals is offensive as implying the view either (a) that animals are a form of property, or (b) the products of human ingenuity or inventions. Whereas the second contention seems to us a serious worry, the first one involves, so we believe, a certain confusion.

5.13 In UK law animals are already a form of property since they can be bought, sold and stolen, and are treated as 'goods' under the provisions of the Sale of Goods Act 1979 – indeed it is the fact that animals are owned which is the basis for the attribution of various duties to their owners in relation to their welfare. Now, just as laws relating to theft protect a certain property interest, so patents are a means of protecting a different form of property interest (an interest in so-called intellectual property), and there is no reason why intellectual property should not reside in animals. It is, after all, human ingenuity or invention which is responsible for the existence of a genetically modified animal in just the form which it has. Thus the extension of patenting to genetically modified animals does not introduce a new doctrine as regards the status of animals as property, but is a logical extension of existing practice. Nor do we find this practice offensive as such, so long as it is understood (as in UK law) that a property interest is not a licence to treat what one owns in any manner one sees fit. There is nothing incompatible then, in recognizing a property interest in animals, genetically modified or otherwise, at the same time as one requires of their owners conformity to a rigorous welfare regime.

5.14 What of the claim, however, that the patenting of animals commits us to the view that they are simply the products of human ingenuity? The entire Committee recognizes the danger, as we have said in previous chapters, that the increasing element of technological intervention in the breeding of farm animals may contribute to a mentality which views animals as no more than industrial commodities, and ignores the fact that however they are produced, they deserve to be treated as living beings with their own natural worth. We differ, nonetheless, on the question as to whether the patenting of animals commits us to this view which we all alike reject.

5.15 Some members of the Committee take the view that the very claiming of a patent on an animal can only be understood as asserting that the animal is an invention. Patents claimed on a gene construct or on the process of modification or insertion of a gene, so it could be argued, may satisfy the formal criteria for patenting, and in particular may be thought of as inventions or as novel. But genetically modified animals are not inventions. Genetic modification of animals is successful primarily in virtue of the inherent capacity of a living animal to integrate what is inserted. Gene constructs incompatible with a living form will not be successfully integrated, and will not result in a genetically modified animal. Thus the successful integration of a gene construct should be recognized as an *essentially biological process*. Since, then, what produces a genetically modified animal are powers and processes natural to that kind of animal, any resultant animal should not be the subject of a patent.

5.16 A majority of the Committee takes another view, and does not believe that the patenting of a transgenic animal needs to be understood as asserting that an animal is simply an invention – what it asserts is that the *particular form* which the animal has and which does not occur in nature, is properly regarded as a creation of human ingenuity, and thus as capable of being patented. A properly drawn patent will claim an interest, after all, in a specific form or type, not an interest in the unmodified animal as such. Furthermore, any protection a patent affords will lapse after 20 years – this seems to us symbolically significant and a further safeguard against the fostering of the view that animals are simply artefacts. The limited degree of protection which patents provide can be understood as implying a recognition of the fact that human ingenuity, in relation to animals or anything else, does not create from nothing, but only modifies existing materials or beings, be they animate or inanimate – for this reason it can appropriately claim only a limited interest in what is, only in a very limited sense, a product of human ingenuity.

5.17 Another argument against the extension of patenting to animals is put by those who oppose all experiments on animals or who, whilst not opposing all experimentation, believe that genetic modification will, in general, be harmful to animal welfare. They argue that a denial of the possibility of obtaining patents will be an effective means of discouraging such work.

5.18 It is clearly the case that the denial of patent protection in relation to animals would discourage this work – the biotechnology industry would not be arguing vigorously for the right to patent work in this field if this were not so. The denial of patent protection to the advances made in this area of biotechnology would not, however, eliminate it altogether, since some would probably think it worthwhile to pursue such work in the hope of protecting their investment by secrecy. Nonetheless, if the Committee were persuaded that genetic modification was likely to be seriously harmful to animal welfare, there would be a case for denying any patents in this sphere.

5.19 Were we to take this view we would not be inclined to give any weight to the considerations which some have advanced relating to the economic interests of the UK in particular and the EU in general, claiming that anything less than complete openness in the matter of patents would give the UK or the EU a competitive disadvantage. As we have already remarked, the abolition of child labour in Victorian Britain gave other nations a competitive edge, but where a practice is plainly wrong such considerations can have no weight.

5.20 We do not, however, take the view that the genetic modification of animals as such can be so regarded, as we have explained in the chapters relating to intrinsic concerns and animal welfare. Some instances of genetic modification seem to us acceptable, whereas others are objectionable. Thus there is reason, we believe, for denying the protection afforded by patents only to some work in this field, and the mechanism for granting patents should be capable of discriminating between the different cases. This contention will not satisfy opponents of all animal experimentation, but, as we have explained in Chapter 2, we have accepted the view which is widely held and is expressed in current legislation, that the humane use of animals is morally acceptable.

5.21 Mention must be made of the concern that patenting of animals will allow the concentration of the ownership of genetically superior stock in fewer and fewer hands, to the detriment of Third World farmers and perhaps also to the detriment of smaller farming enterprises in general. As we acknowledge in the chapter of this report relating to socio-economic concerns, it is difficult to predict with any accuracy the general consequences for farming of developments in this field. What seems certain is that large multinational companies will be most capable of funding the sort of research which will lead to patentable developments, and that they will seek a return on their investment by persuading farmers throughout the world of the benefits of these developments. It does not follow, however, that Third World farmers or small farmers in general will be disadvantaged by these developments. Farmers will go on using existing animal varieties as well as animals whose performance is improved by more traditional breeding programmes and by AI and embryo transfer. Animals so produced cannot be patented, and farmers will only seek to use patented stock if they expect the gain in so doing to offset the cost.

5.22 Cause for concern on this issue is heightened, however, by the breadth of certain patent claims which have been successful in other jurisdictions. The US Patent Office, for example, has granted a patent on all genetically modified cotton to a company called Agracetus. The grant is controversial and is likely to be challenged, and it is far from certain that a similar claim would be successful if made to the European Patents Office. Nonetheless, the obvious ambition of biotechnology companies to claim the widest protection possible for their innovations may pose a threat to the small producer and we recommend that relevant Ministers monitor developments. We do not believe, however, that these concerns are sufficiently strong to warrant opposition to the patenting of living material altogether.

5.23 It has been put to us that the patenting of animals risks unfairness to the small producer in a different regard. It is possible, in principle, that an animal which has been produced by conventional breeding could, as a result of some chance mutation, be identical with one which is the result of genetic modification and protected by a variety of patents. In such a case the farmer might be prevented from breeding from the animal, even though in obtaining it there has been no improper reliance on the research work which led to the patents.

5.24 In reply it can be said that the chances of such an occurrence are vanishingly small and again, do not provide sufficient grounds for opposing altogether the patenting of animals.

5.25 The arguments of those on the other side, who claim that patents should be granted in relation to work on animals without any reference to considerations other than formal ones, we have for the most part addressed already. We would add, however, that their arguments have a force in the context of the situation in the UK which they lack when we consider the position in Europe. If the ASPA ensures, at least to some extent, that animals whose welfare is seriously threatened by genetic modification cannot be released for commercial applications, then it may seem unnecessary to place essentially the same obstacle on such genetic modification but in the context of patent legislation. However, as the extremely helpful Report of the House of Lords' Select Committee on the European Communities[12] put it, "we are not in a position to say that regulation of laboratory experiments and control of exploitation of biotechnology is in itself adequate, since we cannot examine whether the regulation of biotechnological experiments by national, international and Community legislation is comprehensive in its scope and uniformly and adequately applied . . . Patents are certainly not, and should not be, the principal way in which States control research, development and application of biotechnological inventions, but consideration of the application for a patent provides another opportunity for ethical issues to be addressed. We have come to the conclusion therefore that ethical criteria can properly be included in this Directive."

5.26 If, however, the EU Directive represents, as we believe, a good way forward in general, questions may nonetheless be raised about its details. As we have said, the Directive specifically clarifies the general exclusion from patentability of inventions contrary to public policy or morality, by stating that amongst such unpatentable inventions are "processes for modifying the genetic identity of animals which are likely to cause them suffering or physical handicaps without any substantial benefit to man or animals, and animals resulting from such processes." Does this represent a workable and appropriate provision?

---

[12] House of Lords Select Committee on the European Communities, 4th Report, 1993–94, HL Paper 28, *Patent Protection for Biotechnological Inventions*, HMSO, 1994.

5.27   As to the workability of such a provision, we note the concern of the House of Lords' Select Committee that patent examiners faced with too wide a remit are likely to find their decisions challenged in a "proliferation of litigation". "The objective should be legislation which itself provides specific guidance . . . The specific matters to be excluded from possible patent protection should be clearly identified and it should be specified that 'publication or exploitation' of these matters once established 'would be contrary to public policy or morality' without argument as to whether they would otherwise be justified."

5.28   Unfortunately Article 2.3(c), to which we have already referred, does not seem to meet this standard, since it leaves a great deal open to question. This point will become clear if we turn to the matter of the appropriateness of this provision as a means of ensuring the non-patentability of plainly objectionable genetic modifications. Concern will naturally attach to the qualification "without any substantial benefit to man or animal". If patents will be granted for modifications which cause suffering or physical handicaps providing there is substantial benefit, does this provision, it might be asked, constitute a significant ethical hurdle, since it is unlikely that a patent would be sought if there were no such benefit?

5.29   The assessment of what constitutes substantial benefit will be a matter for patent examiners and decisions will be open to challenge. The sort of reasoning which is likely to be considered relevant is illustrated by two cases dealt with under existing legislation, where one application was granted and the other turned down. A patent for the oncomouse (a mouse made susceptible to cancer by genetic modification) was granted by the United States Patent and Trademark Office in April 1988. A European patent was applied for and granted by the European Patent Office (EPO), but that patent is still the subject of opposition proceedings, mostly on moral grounds. In granting the patent the EPO weighed the possible benefits to mankind and the fact that fewer animals might be needed for experiments in future given this modification, against the harm to the animals. A patent was not granted however, for the so-called Upjohn mouse, which was genetically modified so as to be suitable for work on baldness.

5.30   Given the general approach we have outlined in Chapter 2, it will be clear that we do not regard such a consequentialist approach as providing a sufficient test of the ethical acceptability of human action in relation to animals. That is to say, since we have objected to the production of animals whose natural integrity has been damaged by genetic modification, irrespective of welfare considerations and regardless of any alleged benefits which might result, we would prefer the draft Directive to preclude the patenting of such constructs, processes or animal types. Nonetheless, the draft Directive goes some way towards protecting animals from unwarranted use of genetic modification.

Conclusion

5.31   The question we have considered is whether work relating to the genetic modifications of animals should be denied the possibility of gaining protection by patent. Such work may be protected by patents relating to the gene construct, the process of modification, the techniques for inserting modified genes, or to the modified animal itself. Some members of the Committee take the view that patents should not be granted on transgenic animals for the reasons that we have explained. The majority does not regard such patents as objectionable in principle. All of us accept that the case for denying protection by patents to all work relating to genetic modification of animals is not finally persuasive, but believe that a moral criterion does have a proper place in the consideration of patent applications. The draft EU Directive has such a criterion and though we have reservations about this Directive, we nonetheless recommend that

Ministers support it as it relates to animals. Though it may need revision at a later stage, the Directive establishes an important principle which will serve to protect animals to some extent, and for that reason is to be welcomed.

47

CHAPTER 6: GENETIC MODIFICATION AND ENVIRONMENTAL RISKS

Introduction

6.1 There are many examples of domesticated or non-native species becoming established in the wild. In some cases, a species which has been introduced in the past may now be deemed worthy of conservation as a valued part of the local fauna; the dormouse, introduced by the Romans, is such an example. In other cases, however, the introduction has had a major ecological impact which has been unwelcome. An example is the grey squirrel which has displaced the red squirrel in much of England, and further afield, the rabbit, which was introduced into Australia and has had a significant effect on the local ecology. It is a matter for debate whether the introduction of a genetically modified organism to the environment is strictly comparable with the introduction of an exotic species. Nonetheless, there is very understandable public concern that genetic modification may produce organisms capable of having a similar or greater effect if they are released into the wild, and that some of these introductions may be harmful.

6.2 The other breeding techniques which we are considering do not cause the same concern – they are, after all, essentially methods to speed up what could, at least in principle, be achieved by traditional selective breeding. Offspring bred by these techniques may be genetically 'superior' to the parent lines, and may have certain quite different characteristics from their more distant forebears; it is possible, then, that in virtue of these new characteristics they might pose a greater risk to the environment than 'unimproved' stock. Nonetheless, they remain essentially similar to other animals of the same species bred by traditional methods, and are unlikely to threaten the environment in a radically new way or to a radically new degree.

6.3 But can the same be said of genetically modified organisms (GMOs)? Modification of the genome, perhaps by the introduction of genes from unrelated species, may produce an organism unlike any which currently exists or could conceivably be produced by traditional selective breeding even when such traditional breeding is enhanced by AI, embryo transfer, and so on. A transgenic animal could have been modified so as to make it, unlike other members of the species, resistant to certain diseases for instance, or tolerant of severe conditions. Such an animal, if released into the wild (whether deliberately or inadvertently), could pose particular risks to the environment.

6.4 Some of those who replied to our consultation letter were resistant to the idea that this aspect of genetic modification raises *ethical* issues. The resistance to this idea arises from a confusion. Of course the assessment of the nature of the risk, if any, which a genetically modified organism poses if released is a *scientific* matter. Similarly the question as to how to minimize or remove the risk so posed will best be answered by those who have the expertise needed to understand the characteristics of the modified organism, its likely effect on the environment, and so on. There is, however, an *ethical* question which remains when the scientific questions have all been answered, and that is whether any risk which is posed, either by the creation or release of a genetically modified organism, is one which ought to be accepted. This question is obviously one about which there will be disagreement; to take an example, some will think that if the release of a particular genetically modified organism brought with it the 25% chance of the loss of another species, the release would be wrong, no matter what. For others it would depend on their view of the threatened species, or perhaps on the nature of the benefits which the release of the organism offers. But whatever view is taken of the matter, it is plain that the questions under discussion are finally ethical (i.e., questions about what one values) and that anyone who declares certain risks acceptable or unacceptable has, even if unconsciously, a set of values on which that declaration is based.

6.5 The problem of the release of GMOs has been discussed by the Royal Commission on Environmental Pollution Thirteenth Report[13] and though it may not be widely known, the UK and the European Community already have in place detailed regulations relating to the containment and release of GMOs. Our task has been to consider and comment on the adequacy of these regulations in the light of the ethical concern we have identified. It is necessary to begin, however, by saying something more about the risks which GMOs may cause to the environment. Without an understanding of these risks it is impossible to consider the adequacy of the regulations which are in place.

What risks are posed by genetic modification?

6.6 The modification of the genome, and particularly the introduction of new genes from unrelated species into a viable organism, capable of developing and surviving in the wild and passing those genes on to later generations may pose risks to the environment. Whether there is such a risk, and its seriousness, is critically determined by at least three factors: the animal which is subject to the modification, the nature of the modification and the method by which the modification is made. We shall comment on each in turn.

6.7 Some animals which may be subject to genetic modifications, such as deer and fish, are difficult to contain. Others, such as cattle, are more easily confined. Where a GMO does escape, its chances of establishing itself are greater if, again like deer and certain fish, it has close relatives living in the environment with which is can mate and so pass on the modification. This might be by mating with members of the same species, or by hybridization with closely allied species. In either case the new strain is more likely to establish itself as a competitor within the environment if the possibility of inter-breeding exists.

6.8 The risk of a GMO escaping and establishing itself in the wild is partly determined by the nature of the animal and by the existence of potential mates, but the potential risk posed is also determined by the nature of the modification. If the modification confers no particular advantage on the animal in the wild, the fact that it is genetically modified is of no consequence – it will have no better chance of establishing itself and so of affecting the environment than any other unmodified animal which is released. Suppose, for example, a deer were modified so as to produce human proteins in its milk for pharmaceutical use: it would most likely gain no competitive advantage by this modification whereas a fish modified so as to incorporate a copy gene from a flounder and thereby to possess greater tolerance to cold, would gain an advantage, and so might well have a better chance of establishing itself in virtue of the modification. But suppose the modified deer did in fact gain some advantage and establish itself. Whether the risk of its release into the environment is a risk of *harm* is another matter, and again the nature of the modification would be crucial. For even if it colonized the entire population of deer of the same species, so that all wild deer now produced this human protein, it is not clear that this change would be a harmful one – that would depend on whether there are further ecological implications.

6.9 The degree and seriousness of the risk posed by a GMO is determined by the nature of the animal modified, and by the nature of the modification; but the method by

---

[13] Royal Commission on Environmental Pollution, *Thirteenth Report: The Release of Genetically Engineered Organisms to the Environment*, Cm 720, HMSO, 1989.

which an organism has been modified is also a factor in the equation. Some modifications involve the use of retroviruses to create the new genetic construct. Retroviruses replicate in the host genome after introducing their genetic material at random sites in the host chromosome. Recombination between the integrated virus and sequences already present in the chromosome may occur. This can result in chromosomal material from the host nucleus becoming incorporated into the virus, whether from the host's own genetic material or from other viruses integrated into the host. This causes changes to the pathogenicity of the retrovirus, including its oncogenicity and tissue specificity.

6.10 It can be seen then, that whether and to what degree a genetically modified animal poses a risk of harm to the environment is a matter of some complexity. It is wrong to assume that a GMO threatens the environment simply because it is modified: to establish the exact nature of the threat, if any, requires a careful consideration of a variety of factors.

Addressing risks posed by genetic modification

6.11 From the beginning of work in this area, those engaged in it have been aware of the risks it may pose; indeed in 1975 at Asilomar in California, pioneering scientists in the new field met and agreed a moratorium on some aspects of their work, and also that mechanisms should be evolved to ensure that the potential risks of what they were doing would be adequately addressed. Governments have responded by issuing guidelines or imposing regulations, and have generally favoured a cautious approach. Such an approach is appropriate with any new technology, but is particularly so in this case where reliable assessments of the risks posed by genetic modification only become available as the science itself develops.

6.12 In Britain, a regulatory structure has been in place for some years to ensure that account is taken of the risks which genetic modification may pose. This structure has two main elements. One set of regulations under the Health and Safety etc. Act 1974 governs the contained use of GMOs. A second set of regulations under the Environmental Protection Act 1990 (EPA 90) governs the deliberate release of the GMOs including genetically modified animals. These Regulations implement two European Community Directives on contained use and deliberate release: 90/219/EEC and 90/220/EEC.

Contained use

6.13 The Genetically Modified Organisms (Contained Use) Regulations 1992 go beyond the requirements of Directive 90/219, which relate only to genetically modified micro-organisms, and cover all GMOs. They lay down conditions relating to the culture, storage, use, transportation, destruction or disposal of such organisms. The environmental risks associated with work with larger organisms are covered separately by section 108(1)(a) of the Environmental Protection Act together with the Genetically Modified Organisms (Contained Use) Regulations 1993. These require an assessment of the environmental risks to be undertaken and made available for inspection, except where a marketing consent (see next section) has already been obtained for the organism in question. Between them both sets of legislation provide cover for risks to human health and the environment arising from contained use of farm animals.

6.14 The Contained Use Regulations cover laboratory operations, the housing and/or breeding of modified animals in animal houses, and the keeping of modified farm animals restrained by appropriate fencing. Before premises can be used, the Health and Safety Executive (HSE) has to be notified. When the animal is considered to be as safe as its parent these Regulations require an annual retrospective return of the total number of risk assessments, a statement as to whether the activities are to be continued and any changes to the particulars already notified to the HSE. If the contained animal were not as safe to human health as its parent (for example, a honey bee modified to withstand cold and which had a much more poisonous sting), any activity involving the animal has to be notified to the HSE 60 days in advance (or less if the HSE agree). The activity may proceed at the end of this period subject to any extra conditions deemed necessary by the HSE. In cases of doubt or difficulty the HSE may seek advice from the Advisory Committee on Genetic Modification (ACGM).

6.15 Guidance on the risk assessment of contained use of transgenic animals is given in ACGM/HSE/Note 9[14]. This Note dates from 1989 and deals with problems posed by escape into the environment and the possible environment impact of genetically modified animals. The Note is to be amended to bring it in to line with current regulations, but the principle which it establishes is that the precautions to be taken in contained use of a genetically modified animal must relate to the risk which its release poses. The Note addresses in particular the risks posed by the use of retroviruses, and to date such use has been strictly controlled by the Home Office Inspectorate and the HSE to ensure that retroviruses are replication defective, incapable of mobilization and safe.

Deliberate release of GMOs

6.16 Part VI of EPA 90 and the Genetically Modified Organisms (Deliberate Release) Regulations 1992 (as amended) implement EC Directive 90/220. These regulations require the Secretary of State for the Environment's consent to the deliberate *release* of GMOs for research and development purposes. For the *marketing* of a GMO they require consent from the Secretary of State or the competent authority of another Member State of the European Union. It should be noted that no application to release or market vertebrate GMOs has yet been made to any competent authorities in Member States, so far as we are aware.

6.17 The principal duty imposed by the EPA 90 on the Secretary of State in giving his consent to a deliberate release or marketing application, and in agreeing or disagreeing to a consent to market being given in another Member State, is to prevent or minimize any damage to the environment which may arise from the release from human control of GMOs. Damage to the environment is defined as causing harm to the living organisms supported by the environment. Harm is in turn specified as harm to the health of humans or other living organisms or other interference with the ecological systems of which they form part and, in the case of man, includes offence caused to any of his senses or harm to his property.

6.18 When an application is made for consent to release a GMO, the Regulations require that the application be supported by a considerable amount of information as well as an environmental risk assessment. All such applications are considered by the ACRE which advises the appropriate Secretaries of State and other Ministers[15]. Any application for consent to market a GMO is also subject to consideration by the competent authorities of all other EU Member States. In the event of any objections, the application is considered by a committee of Member States which can accept the application by a Qualified Majority Vote, or the application has to be considered by the Council of Ministers. A consent to market a GMO applies in all EU Member States. Consents to release or market may be subject to conditions.

---

14 ACGM/HSE Note 9, *Guidance on work with transgenic animals*, ACGM Secretariat, HSE 1989.
15 The Secretary of State for the Environment, the Secretary of State for Scotland, the Secretary of State for Wales, the Minister of Agriculture, Fisheries and Food, and the Health and Safety Commission.

6.19 In advising the Secretary of State on a proposed release or marketing of a GMO, ACRE takes a very broad view of the possible impact which a GMO may have on the environment. As required by the Environmental Protection Act, consideration is given to the full range of identifiable and possible adverse effects that may arise from a release of a GMO. If ACRE determines that the release or marketing does pose a risk of harm to the environment or to human health, and that no adequate measures are proposed to deal with the risk, then the Secretary of State would be advised that consent should not be given or that additional conditions should be imposed on any consent.

6.20 What is meant by a risk in this context? Most, if not all, human activities bring with them certain risks, and hardly any could be considered risk free. A walk down a street brings with it the risk of being hit by a meteor, but this would not normally enter into one's calculations about the safety of so doing. In considering whether a release poses a risk, the Secretary of State obviously exercises a discretion which cannot be defined, but the concern of those who advise him must relate to risks which can reasonably be anticipated, rather than to all those which can be imagined. There is, of course, a real problem in the anticipation of ecological effects stemming from a release and in some cases this is very speculative – for this reason the continued monitoring of some releases is appropriate. Those who are releasing or marketing GMOs have a comtinuing duty of care under Section 112(5) of EPA 90 with regard to any risks.

6.21 It is important to point out that the Secretary of State is not required by the Act to prevent all change in the environment, only to prevent or minimize damage to it. The environment as we know it is in part the result of human activity and not all future changes can be considered as damaging. Thus, to take a hypothetical example: suppose a genetically modified deer gained a very small competitive advantage by a modification, such that it could be anticipated that its introduction into the wild would cause a very modest increase in the numbers of the present population which would have no further ecological impact. And suppose that the present population do not themselves constitute a nuisance. Then the release of the GMO will probably be considered not to cause harm to the environment, in which case ACRE would not raise objections to consent being given. It can be seen, however, that the interpretation of the word 'damage' in the Environmental Protection Act is a crucial matter.

The importation of GMOs

6.22 Some concerns were expressed to us in regard to the importation of GMOs. Would it be possible for genetically modified farm animals to be imported and released in disregard of the regulations? We understand that imports from other Member States of the European Union or members of the European Economic Area (EEA) would either be under the contained use regulations or have been subject to a marketing consent to which the UK would be party. Imports from other countries other than the EU/EEA countries would also be subject to contained-use rules unless and until consent was given for release or marketing.

6.23 In addition it can be pointed out that non-EU/EEA countries are likely to have controls on GMOs similar to those which obtain here. The principles of risk assessment of GMOs have, in general terms, been agreed by those member nations of the Organisation for Economic Co-operation and Development where most of this work is taking place (i.e. in the developed countries). Other bodies, such as the FAO and UNEP, are seeking to ensure that less developed countries are able to have the protection of a similar level of control.

6.24 However, the very nature of GMOs, and in particular fish, means that we cannot always rely on the formal apparatus of border controls to prevent introduction of

GMOs. We note that Agenda 21 as agreed at the Rio Summit in 1992 called for all countries to introduce adequate controls for GMOs. The Convention on Biological Diversity, also concluded at Rio, called on contracting parties to consider the need for, and modalities of, a protocol to regulate GMOs internationally. Finally, we understand that the Netherlands and United Kingdom governments have drawn up proposed guidelines to provide a framework for such controls. It is not appropriate for this Committee to take a view as to means by which an international consensus on the control of GMOs can be reached, but we recommend that the Government continue to support international understanding, harmonization and co-operation in this area.

Discussion

6.25 The need to address the risks posed by the contained use and release of GMOs will increase considerably in the years ahead. The initial application of the technique of genetic modification is likely to be for the production of high value stock for pharmaceutical purposes, either to provide human therapeutic agents or, more ambitiously, organs suitable for transplantation to humans. Animals bred for these purposes will be kept in protected, contained-use facilities because of their high value. As the development of modification techniques makes them more efficient, however, and as the current genetic mapping programmes reveal more about different species' genomes, genetically modified animals will come to have a place in agriculture in general and applications for consents for their release will become commonplace.

6.26 The risks to the environment posed by the release of genetically modified animals should be taken seriously, and we believe that the current regulations do just that. The principle underlying these regulations for the contained use and release of GMOs is that any potential environmental impact needs to be considered, and that where there is a risk of harm, the use of a GMO should be subject to stringent requirements and either not released or released with sufficient measures taken to ensure that the risk is managed satisfactorily. We note that the legislation does not invite the Secretary of State to weigh possible risks against potential benefits in giving consent to releases, but lays upon him a duty to prevent or minimize harm. In fulfilling this obligation, the Secretary of State is, as we have pointed out, advised by ACRE. The approach of ACRE has been to draw on its own expertise and relevant scientific data, as well as on the fruits of research commissioned by Government into specific aspects of risk. Early releases of plants have been subject to heavily restrictive conditions, either volunteered by the applicants or imposed, to ensure that any risks are adequately managed. As releases have proceeded and more has become known (as a result of the releases or from relevant research projects), these conditions have been relaxed. There is also a public register of GMOs which makes available information on release and marketing applications including the risk assessment, as well as ACRE's advice to the Secretary of State. An open approach is essential if public concern about the new technology is to be allayed. We therefore recommend that ACRE should continue to scrutinize applications for release or marketing of GMOs on a case-by-case basis and impose restrictive conditions when appropriate until research or experience has provided sufficient data on the impact of releases to allow any relaxation of conditions.

6.27 We note that the EC Directives to which we have referred have been strongly criticized by industry in that they impose a burden on the development of products containing or comprising GMOs. We do not agree that there should be no controls on the release of genetically modified animals for farming. However, industry also points to existing legislation which provides for the approval of certain products on grounds which include the assessment of safety and argue that approval under this legislation should be sufficient guarantee. The Government and the European Commission have accepted this point and Directive 90/220 contains specific provision so that, where product legislation provides for an environment risk assessment of GMOs, such GMOs do not need to be separately considered under Directive 90/220. We welcome the fact that in bringing forward amendments to product legislation to allow for an environmental risk assessment for GMOs, the need to assess the risk posed by a GMO has been fully recognized and that the assessment will be no less rigorous than under present regulations.

Conclusion

6.28 Since the release of GMOs may pose environmental risks, it is right that their release be subject to careful consideration. We have examined the regulations which are in place and have found that they are quite properly designed to ensure that GMOs are securely contained and that where consent is sought for their release or marketing, the risks are carefully assessed and consent given only if there is no real threat of harm to the environment. We endorse the cautious, case-by-case approach which the regulations enshrine and believe that, properly enforced, they offer an appropriately high level of protection to the environment without placing improper constraints on industry. Indeed, we endorse the point forcefully put by the Royal Commission: "the biggest brake on the environmental application of genetic engineering could result from an inadequately scrutinized release which caused serious damage to human health or to the environment and destroyed public confidence in both the science and the scientist".

CHAPTER 7: THE IMPACT OF BREEDING TECHNIQUES ON GENETIC DIVERSITY

Introduction

7.1 One concern about the emergence of the new technologies relates to the impact they may have on genetic diversity. Genetic diversity may be lost in two distinct ways – either by the loss of species, or by the loss of variation within species when breeding centres on fewer and fewer animals. The concern is that the emerging technologies may contribute to the loss of diversity in both respects.

7.2 As regards the loss of species, the new technologies, and in particular genetic modification, may be thought to contribute to the problem in a number of ways. It is possible, for example, that advanced breeding techniques may allow scientists to produce farm animals tolerant of conditions in which they cannot presently thrive. Their introduction to new areas may have a significant ecological impact, and may threaten the existence of other species. This issue has been treated separately in Chapter 6.

7.3 Genetic diversity may be diminished not only by the loss of species, however, but by the loss of variation within species. In farm animals this variation finds readily noticeable expression in the existence of a large number of particular breeds (e.g. there are currently some 800 breeds of cattle worldwide), though it is very important to point out that genetic diversity does not invariably express itself in the outward characteristics which are used to identify and define breeds – thus genetic diversity and the existence of rare breeds should not be confused.

7.4 A number of respondents feared that loss of variation (one aspect of which may be the loss of particular breeds) was already occurring under intense selective breeding, and that it would be accelerated by the use of the new techniques. The concern can be addressed by considering the present situation and the pressures which have led to the loss of rare breeds.

7.5 Breeds of farm animal grew up as a result of several factors. Geographical isolation may have accentuated characteristics of the limited breeding stock to such an extent that an identifiable breed was established. Selective breeding aimed at producing an animal especially suitable for local conditions may have had the same effect. However caused, the creation of phenotypically distinctive animals within a species led in the nineteenth century to the formal identification of numerous breeds within Europe and elsewhere.

7.6 The existence of these breeds has, however, been threatened by developments since then. A greatly improved transport system and increased international trade in animals has virtually destroyed the geographical isolation which was a factor in the creation of breeds. Furthermore, changes in animal husbandry have enabled the same animal to be kept in varying environmental conditions. Thus the conditions which favoured the existence of distinctive breeds have disappeared. The farmer wants an animal which is best adapted for the particular purpose for which it is kept, and is now likely to be able to obtain and keep the most favoured of breeds. Thus the older, traditional breeds are being marginalized to hobby or museum status.

7.7 In the UK, for example, though a large number of breeds of sheep (30 or more) remain in commercial use, in cattle, pigs and poultry particularly favoured breeds predominate. The position is as follows.

> (a) In chickens, broilers have been bred from White Cornish and layers from White Leghorn, but in both cases many genes have been incorporated from other breeds; e.g., from Rhode Island Reds to produce brown eggs. Breeding

strategies involve breeding lines for different traits (e.g., growth, growth plus efficiency, growth plus fertility, and so on) which are then crossed to produce hybrids.

(b) In pigs the situation is similar – lines selected for different traits are crossed to produce hybrids. These lines are based on two breeds, Large White (or Yorkshire) and Landrace. Genes have been incorporated from other breeds for specific purposes – e.g., from Durocs to improve meat quality and from Meishans to increase litter size.

(c) In dairy cattle the dominant British Friesian has been replaced over the last ten years by the Holstein-Friesian (from Canada, USA and the Netherlands) which gives a better milk yield. Holstein-Friesians comprise 80-90% of the total dairy herd. Most beef comes from animals crossbred from the dairy herd. Hereford, Aberdeen Angus and Beef Shorthorn were the principal beef breeds, but larger and better muscled continental breeds (Limousin, Charolais, Simmental, etc.) have been replacing them in recent years.

7.8 The concern of some is that the pressures which have led to the loss of rare breeds are pressures which may lead to a narrowing of the genetic base even within favoured lines. Within such lines some animals are preferred to others, and AI, embryo transfer and cloning enable breeders to overcome the physical constraints on the exploitation of these animals.

7.9 Supposing for the moment that the emerging technologies may indeed contribute to a loss of genetic diversity, why should this be a matter for concern? Three related contentions have been put in submissions to the Committee and we shall address each in turn.

(a) Increased homogeneity could lead to problems associated with inbreeding such as increased susceptibility to disease or hereditary abnormality. Further, the loss of diversity in general, and of certain breeds in particular, not only threatens the health of farm animals, but also the very purposes for which they are kept, since it is a loss of the genetic resources which may be required to meet future breeding needs, whether to resolve a serious environmental challenge or to maintain the rate of improvement in farm animals.

(b) With the loss of diversity and the narrowing of the genetic base from which farm animals are bred, the power of fewer and fewer breeding companies will increase as they gain control of desirable genotypes.

(c) Whether or not it has the effects mentioned above, if the loss of diversity includes a loss of existing breeds it represents a harm in itself since these breeds are part of our heritage and should be valued for themselves.

Homogeneity

7.10 The narrowing of the genetic base from which animals are bred increases their genetic homogeneity. When, however, does this narrowing become harmful (either as leading to the problems associated with inbreeding, or by reducing genetic resources from which future improvements or developments may come) and is this narrowing likely to occur as a result of the application of the emerging technologies?

7.11 The problems associated with inbreeding are well-known: inbreeding increases homozygosity, and thus increases the incidence of conditions caused by recessive genes. It thus reduces the general fitness of progeny and may produce serious hereditary abnormalities. An increase in homozygosity is also, of course, a loss of genetic variation.

7.12 Genetic variation can be measured in a number of ways: by reference to visual traits such as colour and horns; by reference to quantitative traits such as milk yield or conformation; by reference to blood groups, enzymes and DNA, which can be used as neutral markers; or by reference to particular and important known genes.

7.13 The loss of genetic variation is essentially a function of population size and selection. The larger the population, the more males used and the more equal the family representation, the slower the loss of variation. In relation to quantitative traits, however, selection is an additional factor, serving to reduce the effective population size.

7.14 The loss of variation in a closed population is balanced, however, by genetic mutation and in quite small population sizes, perhaps of no more than 100, a balance between loss of variation and its introduction by mutation is achieved. Variation is also maintained by opening a population to genetic material from independent lines.

7.15 It can be seen then, that the key to maintaining genetic variation is ensuring adequate population sizes and the existence of independent populations. In relation to the UK's major breeds of poultry, pigs and dairy cattle there exist large and independent populations – in beef cattle and sheep the population sizes are smaller, but quite high enough to ensure the maintenance of genetic variation. This situation does not exist by chance – variation is what breeders use in order to effect improvements in their stock, so the need to maintain diversity has always been well understood. Breeders are also conscious of the fact that the objectives and priorities of breeding programmes change in accordance with consumer preferences. The current demand for leaner animals, for example, may give way to a demand for more intramuscular fat for the sake of taste. The preservation of genetic variety is essential then, not only for the making of improvements in currently valued traits, but also as the means of responding to changes in demand.

7.16 The maintenance of genetic diversity is then relatively straightforward, and the need to maintain this diversity is something of which all breeding companies are aware. The application of the new technologies does not, as such, threaten this situation, or render likely a dangerous narrowing of the genetic base from which farm animals are bred. Indeed most of the new technologies we are considering offer highly effective means of introducing genetic diversity. Take AI and embryo transfer for example. Prior to their introduction, effective population sizes and the number of independent lines were quite small because a small group of elite breeders provided stock via a pyramid structure. The introduction of AI and embryo transfer has introduced variation by enabling the easy use of material from other populations – thus, to take a specific instance, variation in quantitative traits such as milk yield was introduced in the British population of Friesians with the importation of North American Holstein animals. Likewise genetic modification is a means of increasing variation in a population, perhaps by the introduction of genes from another species, rather than a means of reducing it. It is, like mutation, a route to increased diversity.

7.17 It is, of course, the case that these technologies will only have beneficial effects on genetic diversity if used in populations of a sufficient size to ensure that variation is maintained. For example, though the Holstein-Friesians population is large, the number of males used has rapidly decreased over the years through the use of AI, and UK breeders will now have to monitor the situation carefully. But exactly the same difficulties can arise in the practice of traditional selective breeding, which must respect the familiar considerations to which we have referred.

7.18 Cloning, if it ever becomes commercially viable for livestock production, would lead to a reduction in the variation to be found in a typical herd of cattle, for example, supposing that farmers found it advantageous to keep a large number of genetically identical animals. But such animals kept for production purposes would not provide the breeding population, which would be maintained with the diversity which is essential to future improvements.

Control of genetic resources

7.19 It is contended by some that since the new technologies will contribute to a loss of diversity and to the narrowing of the genetic base from which farm animals are bred, the power of breeding companies will increase as they gain control of desirable genotypes. For reasons we have explained, we do not accept that the emerging technologies will contribute to a narrowing of the genetic base. Nor is it clear that these new technologies, with or without that effect, will increase the power of large breeding companies. Of course, the effect which the introduction of the new technologies will have on the pattern of commercial animal breeding is extremely difficult to predict. In the poultry industry the current position is that a few major breeding companies supply the market world-wide; this is not however due to the introduction of new technologies. In relation to pigs and cattle, where AI and, in cattle, embryo transfer are increasingly important, large breeding companies also have a significant place in supplying animals to farmers. As selective breeding draws on more sophisticated data analysis, and is supplemented in due course by some of the presently experimental techniques we have been considering, it seems likely that economies of scale will favour these large companies. Since, however, individual producers will continue to have the freedom either to purchase from such a large breeding company, or to breed from their own animals, it is difficult to envisage these companies gaining an unfair advantage over farmers, who will purchase the latest improved stock only if they expect to benefit from so doing.

Loss of breeds

7.20 If the present pattern of farm animal breeding ensures the maintenance of genetic diversity within breeds, is there any reason to be concerned at the loss of breeds? One concern which is often voiced is that rare breeds may contain valuable genetic variation.

7.21 A breed is a population which can be distinguished by its **phenotype**. This might be a matter of visible traits (such as colour, size, etc.) or of production traits (e.g. high fat content in Channel Island milk). It is possible, however, that phenotypic variation between two breeds might be very striking, even though at the level of genotype the two breeds are very close indeed. By the same token, considerable genetic diversity, which does not express itself in obvious external characteristics, could exist within a breed whose members are almost indistinguishable to the eye. It cannot be assumed then, that the loss of rare breeds threatens a significant loss of genetic diversity. Not enough is presently known about the genetic relationships between breeds to say whether or not the differences between them reflect considerable diversity, and the FAO[16] is only now beginning a project which will provide this information. It is a further question whether such diversity, if it exists, is of commercial significance. One reason why rare breeds are disappearing is that, except perhaps in very particular circumstances, they have very few traits which are of real value, lagging far behind popular breeds in terms of performance. It is not beyond doubt that there may be some gene or genes in rare breeds which could be of some future value, such as a gene conferring resistance to disease. But commercial populations possess considerable variation within themselves and many breeders would take the view that

---

[16] Food and Agriculture Organisation of the United Nations

within these populations variation for most traits is still present. And even if a potentially useful gene were identified in a rare breed, the problems of introducing it into commercial stock efficiently, even with techniques used in genetic modification, are considerable – thus a breeder faced with a problem within a parent stock would turn to another improved but independent population before thinking of utilizing an unimproved breed. A difficulty would exist, however, if intense selection according to particular criteria resulted in the virtual eradication of a trait (for example, resistance to sunburn in pigs) in the entire improved population, when at some point in the future, perhaps under different husbandry conditions, such a trait could be vital.

7.22 Whether or not rare breeds offer a source of useful genetic material for future breeding programmes, their disappearance is something which many will properly regret. They represent part of the history of farming, and even if they find no place on the average farm, the continued existence of a variety of breeds in particular localities adds to the richness of the countryside. It does not seem clear, however, that the existence of rare breeds is in actual fact threatened by the emergence of the new technology. Sophisticated selective breeding and improved transport and husbandry were quite sufficient to ensure their displacement, whereas the new technologies may actually be of assistance in their preservation. AI and embryo transfer, for example, make it easier for a breeder to obtain genetic material from the possibly small number of other breeders who maintain a particular rare breed.

7.23 In so far, however, as the degree and value of the diversity within rare breeds and the likely effect of the new technologies on the pattern of commercial breeding are both uncertain, we welcome the recent initiative from the FAO, following the signing of the Rio Convention on Biological Diversity, in sponsoring a project to identify the genetic relationship and variation between and within animal breeds throughout the world.

7.24 The European Union also adopted in June 1994 a regulation on the conservation, characterization, collection and utilization of genetic resources in agriculture (Council Regulation 1467/94). The aim of this regulation is to co-ordinate and promote work in this area by member states, and it gives particular encouragement to programmes to map the genomes of farm animals. Only with the knowledge which the mapping of animal genomes will provide, can an accurate assessment of the present level of diversity between breeds be made, and appropriate action taken.

7.25 Until this work has been completed, it cannot be ruled out that useful genes will be found in presently threatened breeds. We accept, in any case, that such breeds are interesting and valuable in their own right. On either count, it is right that steps should be taken to preserve them. We note that the Government's plan to implement the Convention on Biological Diversity, the UK Biodiversity Action Plan[17], makes a number of references to the conservation of farm species, and we recommend that further consideration be given, by Government, to the need for specific measures to conserve these breeds. Such measures, whether appropriately sponsored by Government or by others, might include for example:

(a) the establishment of a UK register of breeds, to record their numbers and population sizes;

(b) a survey to measure diversity within and between breeds using molecular markers and production traits;

(c) the construction of a biodiversity database (the Government has already accepted the need to establish such a database, and this is mentioned in the UK's Biodiversity Action Plan, paragraph 9.37); and

---

[17] *Biodiversity: The UK Action Plan*, Cm 2428, HMSO 1993, ISBN 0-10-124282-4.

(d) the establishment of a genome bank where gametes and embryos are cryogenically stored for use at a later date to re-introduce genes that have been lost from a population.

Conclusion

7.26 We have taken the view that there is no reason to suppose that the new breeding technologies will have a significant impact on genetic diversity in farm animals. The interest of breeders lies in maintaining variation, and the increasing use of these technologies will not alter that situation nor discourage breeders from acting in such a way as to protect diversity. The loss of rare breeds, which may or may not mean a loss of valuable genetic variation, has been caused by many factors, and cannot be blamed on the new technologies; indeed these new technologies may have a part to play in assisting in the preservation of rare breeds by enabling the effective use of the remaining animals in breeding programmes which will ensure their survival.

CHAPTER 8: SOCIO-ECONOMIC CONSIDERATIONS

Introduction

8.1   Not least amongst the concerns which are aroused by the emerging breeding techniques is a concern as to the effects that they may have on the economic and social life of the community.

8.2   Those who support the widespread application of biotechnology in general, and the new breeding techniques in particular, claim that the farmer and the consumer will both benefit from the new technology. More efficient farms, so it is said, will produce cheaper and more varied food and even such novel products as pharmaceuticals. It is further argued that UK and European agriculture will be able to compete in the more competitive environment which is heralded by the GATT agreement only by taking advantage of the opportunities the new breeding techniques offer. Nor will the advantages accrue only to the developed countries according to the advocates of biotechnology – as the advances in agricultural practices become commonplace in developed countries, so they can be applied in developing countries, playing a part in meeting the various needs of the world's growing population.

8.3   If these are the advantages sought by the application of the new technology, there may also, however, be unexpected disadvantages. These effects may range from the relatively small-scale to the very large scale – some predict that the application of the new technologies may threaten the viability of family farms, whilst others anticipate damage to Third World economies.

8.4   It might be argued that the consideration of the socio-economic impact of these techniques is out of place. In a free market, some would say, economic forces must be left to themselves. To try to anticipate the social effects of commercial developments with a view, perhaps, to their regulation, is to restrain progress and development.

8.5   This point would have more force were it not for the fact of the existence of the interventionist Common Agricultural Policy (CAP) under which the practice of agriculture is regulated in the European Union and which has as one of its aims "to ensure a fair standard of living for the agricultural community" (Article 39.1(b) of the Treaty of Rome). Whether or not it was based on sufficient evidence, the recent proposal to ban the use of Bovine Somatotrophin (BST) within the Community until the end of the milk quota system, partly on the grounds that it would favour large producers, was at least consistent with existing policy.

8.6   If, however, a consideration of the socio-economic impact of the new breeding techniques is appropriate, it is hampered by the sheer complexity of modern societies and economies, which renders any predictions in this area more than a little hazardous. This point cuts two ways, of course – prophets of a golden age as much as prophets of doom stand on unsure ground – but it does not mean that the task of anticipating the unexpected disadvantages of the new technology should simply be abandoned. It means rather that any predictions about the socio-economic impact of the new breeding techniques, either for good or bad, should be made with due caution.

Concerns expressed

8.7   Concern about the effects of the introduction of the new breeding techniques expressed in responses to our consultation letter centred chiefly on four issues. The following views were expressed and we examine each in turn.

(a) The use of the new breeding techniques will simply add to the present over-supply of many agricultural products in the European Union.

(b) Only larger and more intensive units will be able to benefit from the technology which is being developed.

(c) Far from benefiting the Third World, these technologies are largely irrelevant to their needs and may even worsen their economic position.

(d) Since the public is suspicious of biotechnology in general, the widespread introduction of the new breeding techniques may alienate the consumer and threaten public acceptance of all agricultural products.

Will the new techniques simply increase surplus production?

8.8 The first two issues we are considering have also arisen in the debate about whether BST should be approved for use in the European Union. BST is a growth hormone produced by techniques of genetic modification in sufficient quantities to be administered to dairy cows, thereby boosting production of milk. Though it does not as such fall within the scope of our enquiry, it is interesting to note that opposition to the use of BST, as well as raising issues of human and animal safety, centred on the claim that it would add to already excessive production of milk and would encourage the creation of ever larger dairy units.

8.9 As regards levels of production, it was pointed out that even in the context of the CAP, there is an advantage in more efficient production of a product, even if an increase in total production is unwelcome. That is to say, the use of BST would enable whatever is regarded as optimum production to be achieved by fewer cows, with less cost to the farmer and the environment. In this case even this advantage would have an accompanying disadvantage, since the reduction in the surplus population of dairy cattle would, in the short term, add to the beef surplus, but the general point is valid.

8.10 In the case of the novel breeding techniques, it is not as clear as it is with BST that their wider application would have a significant impact on levels of agricultural production, since the aim of breeding programmes is not simply to produce more or larger animals. Selective breeding making use of AI or embryo transfer may aim to produce animals offering better feed conversion, with greater resistance to disease and higher fertility, or more in line with consumer preferences. The intention would not necessarily be to increase production, but to lower the costs of production or to provide a better product. Similarly genetic modification may have these aims, or may be used to produce non-food products.

8.11 If, however, the application of these techniques did add to production, the point made in relation to BST would apply – that even within a framework of production quotas, gains in efficiency in producing at the desired level may be of value to the farmer, the consumer and perhaps to the environment too. It should also be pointed out that the CAP may be subject to reform, and that if the GATT settlement leads to a free market in world agriculture, both gains in efficiency and increases in production could be vital to the competitiveness of European producers.

Will the new techniques benefit only larger units?

8.12 The question as to whether the introduction of BST would help only larger units is hotly contested, and the same debate exists in relation to the new breeding technologies. It is clear, however, that the use of AI has been very general indeed, and that embryo transfer could have the same significance. Of course some of the other more sophisticated techniques, such as cloning and genetic modification, are likely to be used by only the very largest producers, if at all. But even here it is not impossible that animals bred by these means will become widely available and of use to the smaller producer.

8.13 Though it is difficult to anticipate the applications of the new breeding techniques and the effect of these applications on small farmers, it is apparent that there are diverse forces favouring the larger units which would not cease to operate even if the advanced breeding techniques were banned forthwith. To take just one example – the increasing dominance of supermarkets in the retail sector gives an advantage to units of sufficient size to market efficiently to them.

8.14 If, however, the concern about the effect of these techniques on small farmers does not provide sufficient grounds for opposing their introduction altogether, the concern is a significant one and ought not to be ignored. We not that a concern about the position of small farmers is evident in some aspects of the CAP. It is appropriate, therefore, that Government should be sensitive to the impact of the new techniques, when and if they gain a wider use. We therefore welcome the fact that MAFF, together with the Meat and Livestock Commission and Genus (the commercial arm of the former Milk Marketing Board), is already funding a project looking at the various implications of these techniques, including the socio-economic implications. Work on this latter aspect is being carried out by a group at the University of Reading. The work is in its early stages and is based on what can only be assumptions about the likely take-up of the new techniques. Nonetheless this project, and future monitoring of the situation, will provide the sort of information which is vital if Government is to take steps to ensure that smaller and less favoured farms are not seriously disadvantaged by these developments.

Will the new techniques harm the Third World?

8.15 In response to the large claims which have been made for the benefits of biotechnology to the Third World, it is contended by some that biotechnology is largely irrelevant to the needs of the Third World and may actually be harmful.

8.16 It is certainly the case that if the principal need of the Third World is for food to feed its growing population, the meeting of that need is not dependent on the spread of biotechnology in general or of the emerging technologies we are discussing in particular. Since the growing of crops is a vastly more efficient way of generating foodstuffs than is the keeping of animals, changes in the pattern of farming practice might well on their own transform the situation without recourse to these developments.

8.17 That said, however, it remains the case that there are areas of the Third World which, for reasons of climate, topography and so on, are unable to support realistic efforts to grow cultivated crops and so depend heavily on animal products – and in all parts of the Third World animals have an important place as providers of power or as efficient converters of organic waste into meat, milk, eggs and fuel. If the new reproductive technologies could be utilized at a reasonable cost to raise the quality of livestock, the benefit to the local population may be very significant.

8.18 It should be pointed out, furthermore, that agricultural policy and practice in the developed world will continue to have a profound impact on the developing world, and that barring more imaginative solutions to its problems, the Third World will need to share in biotechnology if it is not to fall further behind. At present, First World agricultural surpluses distort world markets as products are exported with heavy subsidies, while home markets are heavily protected. This system leaves Third World farmers unable to compete, sometimes even in supplying food to their own urban populations. With the GATT agreement the situation may change, but if the new breeding techniques render the agricultural producers of the developed countries more efficient, Third World producers may find themselves once more at a disadvantage. In these circumstances, various initiatives presently underway under the aegis of the FAO and the International Agricultural Research Centres of the Consultative Group on International Agricultural Research, are to be welcomed: they aim to provide less developed countries with the necessary infrastructure and knowledge to build up local livestock industries.

8.19 It should be noted that the transfer of these technologies must be appropriate to the needs and circumstances of the receiving society, and ought not to consist of an attempt to transplant European or North American farming techniques. Farming in the developed world demands high inputs and even if sustainable in some contexts, it is unlikely to be sustainable in the Third World. Incremental improvements in local agricultural systems are, however, appropriate, and in such a context the deployment of the techniques we have been discussing may be advantageous to Third World farming.

Will the new techniques alienate the consumer?

8.20 One certainly unlooked for consequence of the application of the new breeding techniques, and in particular of genetic modification, could be the creation of a public suspicion of certain, or perhaps all, agricultural products. Such a climate of suspicion would harm the whole industry and it is significant that in submissions to this Committee and in other surveys, producers of products such as venison and salmon are highly cautious about the introduction of technology which might tarnish the image of their products as providing traditional, natural and wholesome food. Certainly the evidence of resistance in the United States by some consumer groups to the introduction of milk produced from cows treated with BST warrants a degree of caution.

8.21 It is, of course, possible to take the robust view that consumers are primarily interested in price, and will buy products, however produced, so long as they represent good value for money. It would, however, be a considerable gamble for the food industry to base future policy on such a view of the consumer, and it seems more sensible for the industry to consider carefully both the public acceptability of products deriving from the use of the new techniques and the effect that the introduction of such products may have on public confidence in the farming industry.

8.22 We have noted in an earlier chapter that there has been an unfortunate tendency to dismiss public concerns about the new technology as arising from the so-called 'yuk factor'; that is to say, from an emotional and irrational hostility to science and technology which can be expected to disappear as people become used to the idea of genetic modification, etc. We have taken the view that the public concern is often perfectly rational and deserves to be treated seriously, and that public suspicion of the industry will only be deepened by scornful disregard of its concerns.

8.23 In their own interest, therefore, if for no other reason, those engaged in the development and application of agricultural technology should endeavour to be sensitive to public concerns about the new technologies, open to debate with interested parties, and supportive of a reasonable system of regulation, provision of information and labelling. As regards openness to debate, we believe that it would be wise for the industry to consult interested parties as a matter of course as they consider novel and potentially controversial applications of the new technology. As regards regulation, we would make the point that far from being a restraint on trade, the existence of a system of regulation in which there is public confidence is a vital element in ensuring public acceptance of legitimate applications of the new technology.

ANNEX A: CONSULTATION LETTER

To all interested organisations or individuals

14 December 1993

Dear Consultee

THE ETHICAL IMPLICATIONS OF EMERGING TECHNOLOGIES IN THE BREEDING OF FARM ANIMALS

Ministers have set up a committee to consider the ethical issues associated with the breeding of farm animals. Membership of the Committee and its terms of reference are set out at Appendix A to this letter (not attached).

In considering the ethical implications of emerging breeding technologies, the Committee is looking both at the techniques which are already being applied commercially and at those still being developed. The basic and oldest technique is selective breeding, which has become increasingly sophisticated as the science of genetics has evolved. Artificial insemination and embryo transfer are also well established and widely applied techniques for the enhancement of stock. More recently, an increasing understanding of the reproductive cycles of animals has lead to the development of *in vitro* fertilisation, embryo sexing and pre-determination, and cloning. Another novel technique which may have commercial application is the production of animals by genetic modification. These techniques are described more fully in Appendix B to this letter (not attached).

It is, of course, the case that some people object altogether to the keeping of farm animals. It is not, however, within the Committee's remit to review such objections; its purpose is to consider whether and to what extent, if at all, the emerging techniques in the breeding of farm animals are themselves, in particular, a cause for ethical concern. In this regard the Committee has identified a number of issues for consideration:

(i)   the concern that there may be intrinsic objections to some of these techniques, and in particular genetic modification;

(ii)   the effect of these techniques on the welfare of farm animals used for breeding purposes and on their progeny;

(iii) the effect of these techniques on the genetic diversity of farm animals;

(iv)   the risks, if any, to human health or the environment which may arise from use of any of these breeding techniques;

(v)   the use of patent law in respect of advanced breeding techniques for farm animals; and

(vi)   the impact of the use of advanced breeding techniques on the social and economic life of the community as a whole.

The Committee would be grateful to hear your comments as an interested party in order to assist in its consideration of these issues. Such comments would be particularly helpful if they:

–   explained the reasons for any concern held and the specific breeding technique(s) which gives rise to that concern;

- considered whether that concern is addressed by current legislation and welfare codes; and

- offered suggestions as to the means by which that concern might be addressed if existing legislation is inadequate.

The Committee would be grateful for your views *by 11 March 1994*, and for an earlier indication of the main points of your submission if you propose very substantial comments. These should be sent to Mr Paul Kilby, Biotechnology Unit, Ministry of Agriculture, Fisheries and Food, Room 23, 10 Whitehall Place (East Block), London SW1A 2HH (Fax: 071-270 8656).

In order to help inform debate on the issues raised by this consultation document, MAFF intends to make publicly available, at the end of the consultation period, copies of the responses received. The main Departmental Library at 3–8 Whitehall Place, London SW1A 2HH (Tel: 071-270 8000) will supply copies on request to personal callers or telephone inquirers. It will be assumed, therefore, that your response can be made publicly available in this way, unless you indicate that you wish all or part of your response to be excluded from this arrangement.

If you have no objection to your response being made available for public examination in the way described above, please supply an additional copy of your response to this letter.

Yours faithfully

S B Marshall
Secretary to the Committee

Advocates for Animals
Anglican Society for the Welfare of
   Animals
Animal Aid
Animal Biotechnology Cambridge Ltd.
Animal Christian Concern
Animal Concern
Animal Defence Society Ltd.
Animal Health Distributors Association
   (UK) Ltd.
Animal Health Trade Association
Animal Health Trust
Animal Vigilantes
Animal Welfare Foundation
Anjuman-E-Gujarate Muslim Society
Arthur Rank Centre
Asda Stores Ltd.
Associated Artificial Insemination
   Centres
Association of British Muslim Scholars
   of Great Britain
Association of British Pharmaceutical
   Industry
Association of District Councils
Association of Local Authorities
Association of Metropolitan Authorities
Association of Port Health Authorities
Athene Trust
Babraham Institute
Baha'i Community of the UK
Barling, Mr D
BBSRC Centre for Genome Research
BBSRC/MRC Neuropathogenesis Unit
Beauty without Cruelty
Belgian Embassy
Bhartiya Vidya Bhakan
Bill Sykes and Associates Pty. Ltd.
BioIndustry Association
Biotechnology and Biological Sciences
   Research Council (BBSRC)
Board of Deputies of British Jews
Board of Shechita
Board of Social Responsibility of the
   Church of Scotland
Brander, Mrs J
British Angora Goat Society
British Association of Biotechnology
British Association of Sheep
   Contractors
British Chicken Association Ltd.
British Chicken Information Service
British Commercial Rabbit Association

British Council of Churches
British Deer Farmers Association
British Deer Producers Society Ltd.
British Deer Society
British Dietetic Association
British Domesticated Ostrich
   Association
British Egg Association
British Egg Industry Council
British Goat Society
British Goose Producers Association
British Home Stores Ltd.
British Housewives League
British Industrial Biological Research
   Association
British Leather Confederation
British Meat Manufacturers
   Association
British Medical Association
   (Edinburgh)
British Medical Association (London)
British Milksheep Association
British Nutrition Foundation
British Organic Farmers
British Pig Association
British Poultry Breeders and
   Hatcheries Association Ltd.
British Poultry Meat Federation Ltd.
British Rabbit Council
British Retail Consortium
British Sheep Dairying Association
British Society of Animal Production
British Standards Institute
British Trout Association
British Union Conference of the
   Seventh-Day Adventists
British Union for the Abolition of
   Vivisection
British United Turkeys Ltd.
British Veterinary Association
British Veterinary Association (Animal
   Welfare Foundation)
Buddhist Meditation Centre
Buddhist Society
Cambridge University (Animal Welfare)
Campden Food And Drink Research
   Association
Cardinal Cahal Daly
Care for the Wild
Carroll, Ms A
Catholic Bishops' Joint Committee on
   Bio-Ethical Issues
Catholic Study Circle for Animal
   Welfare

Central Council for Agriculture and
    Horticulture Co-operation
Chadwick, Mr J
Charles River UK Ltd.
Chartered Institute of Patent Agents
Cherry Valley Farms Ltd.
Chief Rabbi
Christian Consultive Council for the
    Welfare of Animals
Church of Ireland Board for Social
    Responsibility (NI)
Church of Ireland Diocesan Office
Church of Scotland, Society, Religion
    and Technology Project
Co-operative Union Ltd.
Co-operative Women's Guild
Cobb-Vantress Incorporated
Common Law Institute of Intellectual
    Property
Compassion in World Farming
Consumer Association
Consumer Watch
Consumers in the EC Group (UK)
Cotswold Pig Development Co. Ltd.
Council for Small Industries in Rural
    Areas
Council of Mosques – UK and Eire
Council of Rabbinical Authority
Council of Welsh Districts
Country Landowners Association
Coward, Mr J L
Cranfield Biotechnology Centre
CWS Quality and Consumer Care
D'Arcy Masius Benton and Bowles Ltd.
Dairy Trade Federation
De Montfort University (Dept. of
    Applied Biology)
Deer Liaison Committee
Digest: Food Policy And Legislation
Dixon Smith (Lyons) Ltd.
Domestic Poultry Keepers Federation
Duck Producers Association Ltd.
Dundee Institute of Technology
Earthkind
Eastern Health and Social Services
    Board
Echlin, Dr E P
EMBREX
Environmental Health Briefing
Episcopal Church of Scotland
Ethicon Ltd.
Europe World Society for the Protection
    of Animals
European Islamic Mission
Evangelical Movement of Wales
Express Food Groups Ltd.

Fallows, Dr Stephen J
Farm and Food Society
Farm Animal Welfare Co-ordinating
    Executive
Farm Animal Welfare Council
Farm Animal Welfare Network
Farm Livestock Welfare Advisory
    Group
Farmers' Union of Wales
Federation of Deer Management
    Society
Federation of Synagogues
Food and Drink Federation
Food Commission
Food from Britain
Food Manufacturers Federation
Food Research Institute
Food Safety Advisory Centre
Free Church of Scotland
Free Presbyterian Church of England
Fuller, Mrs G
Gairn, Ms Catherine
Gateway Food Markets Ltd.
General Consumer Council for
    Northern Ireland
General Synod of the Church of
    England
Genetics Forum
Genus
Glaxo Animal Health Ltd.
Goat Producers Association
Godfrey, Mr J
Green Alliance
Greenpeace
Guild of Food Writers
Hampshire Down Sheep Breeders
    Association
Hannah Research Institute
Harrison, Mrs Ruth
Hatchers Poultry
Henry Doubleday Research Association
Highlands and Islands Enterprise
Hindu Centre (London)
Holstein Friesian Society of Great
    Britain and Ireland
Holt, Mrs
Horrox, Mr Nigel
Hoskin, Mr
Hotel Catering and Institutional
    Administration Association
Iceland Frozen Foods PLC
Imam of Woking
Imperial Chemical Industries
Imutran Ltd.
Institute of Animal Health
Institute of Biology

Institute of Environmental Health
  Officers
Institute of Food Science and
  Technology
Institute of Food Science and
  Technology (Edinburgh)
Institute of Food Technologists
Institute of Grassland &
  Environmental Research
Institute of Laboratory Animal
  Techniques
Institute of Terrestrial Ecology (Deer
  Liaison Committee)
Institute of Trading Standards
International Food Information Service
International Supreme Council of Sikhs
Intervet Laboratories Ltd.
Islam and Mosques Council UK
Islamic Circle Organisation
Islamic Cultural Centre
Islamic Education Trust
Islamic Foundation
Islamic Medical Association
Islamic Sharia Council
J Sainsbury PLC
Journal
Karma Kagyu Buddhist Centre
Kempsey, Mr Richard
King, Mr David
Laboratory Animals Breeders
  Association
Lang, Mr Alen
Laying Battery Manufacturers
  Association of Great Britain
Leatherhead Food Research Association
Leeds University
Leicester University
Linzey, The Revd Professor Andrew
Liverpool University (Faculty of
  Veterinary Science)
Livestock Marketing Commission (NI)
  Ltd.
London Board for Shechita
Long, Mr T N
MacDonald, Mrs Jose
Marks and Spencer PLC
Marr, The Revd Peter
Marshall Food Group Ltd.
Masterman-Lister, Mr M J
McDonalds Restaurants Ltd.
Meat and Livestock Commission (Pig
  Breeding Centre)
Methodist Church (The Revd C Eyre)
Methodist Church Division of Social
  Responsibility
Milk Marketing Board

Moredun Research Institute
Muslim Research Institute
Muslim Concern UK
Nabarro Nathanson
National Agricultural Centre
National Association of Breeders'
  Services (UK)
National Association of Local
  Government Officials
National Association of Women's Clubs
National Cattle Breeding Association
National Citizens Advice Bureau
  Council
National Consumer Council
National Council for Shechita Boards
National Council of Women of Great
  Britain
National Dairy Council
National Dairymen's Association
National Equine Welfare Council
National Farmers' Union (London)
National Farmers' Union (Swansea)
National Farmers' Union of Scotland
National Federation of Consumer
  Groups
National Federation of Meat Traders
  Association
National Federation of Stockmen's Club
National Federation of Wholesale
  Poultry Merchants
National Federation of Women's
  Institutes
National Federation of Young Farmers'
  Clubs
National Food Alliance
National Game Dealers Association
National Housewives' Association
National Institute of Poultry
  Husbandry
National Sheep Association
National Society Against Factory
  Farming Ltd.
National Union of Agricultural and
  Allied Workers
National Union of Townswomen's
  Guilds
National Resources Institute
Nature Conservancy Council
Newcastle Food Ltd.
Nicolson, Mr D
Northern Ireland Agriculture
  Producers' Association
Northern Ireland Dairy Trade
  Federation
Northern Ireland Egg Merchants
  Association

Northern Ireland Federation of Meat Traders

Northern Ireland Master Butchers Association

Northern Ireland Meat Exporters Association

Northern Ireland Poultry Federation

Northern Pig Development Co. Ltd.

Nu-Swift International

Nuffield Council of Bioethics

Office of the Chief Rabbi

Orban Association

Owen, Mrs S

Oxford University (Dept. of Biological Sciences)

Pandariman Trust

Parents for Safe Food

Peel Holroyd and Associates

Pig Improvement Company

Pig Veterinary Society

Pigs Marketing Board

PPL Therapeutics (Scotland) Ltd.

Presbyterian Church in Ireland

Presbyterian Church of Wales

Procter and Gamble Ltd.

Professional Herd Persons Society

Public Health Laboratory Service Board

Quaker Concern for Animals

Quaker Social Responsibility and Education

Rabbinical Authority of the Union of Orthodox Hebrew Congregations

Radha Krishna Temple

Rare Breeds Survival Trust

Reading University

Red Deer Commission

Representative Body of the Church in Wales

Research Engineering Ltd.

Roger, Mr P A

Roslin Institute

Ross Breeders Ltd.

Rowett Research Institute

Rowland Sallingbury Casey

Royal Agricultural College

Royal Agricultural Society of England

Royal Association of British Dairy Farmers

Royal College Of Physicians (Faculty Of Community Medicine)

Royal College of Veterinary Surgeons

Royal Highland and Agricultural Society of Scotland

Royal Society (Animal Experiments Committee)

Royal Society for Health

Royal Society for the Prevention of Cruelty to Animals

Royal Society of Edinburgh

Royal Veterinary College

Royal Victoria Hospital

Royal Welsh Agricultural Society Ltd.

Rural Agricultural and Allied Workers National Trade Group

Safeway Foodstores Ltd.

Scottish Agricultural College

Scottish Association of Meat Wholesalers

Scottish Centre for Animal Welfare Sciences

Scottish Consumer Council

Scottish Egg Trade Association

Scottish Natural Heritage

Scottish Salmon Growers Association

Scottish Society for the Prevention of Cruelty to Animals

Scottish Womens' Rural Institute

Senior Advisory Group on Biotechnology (SAGB)

Shanks, Miss K

Sheep Veterinary Society

Shuttleworth College

Simmons and Simmons

Society for the Reformation of Muslims in the UK

Society for the Study of Animal Breeding

Spring, Mr D G

St Ivel Technical Centre

Stirling University

Surate Muslim Khalifa Society

Sussex University

Tesco Stores Ltd.

Townswomen's Guilds

Trades Union Congress

UK Action Committee for Islamic Affairs

UK Association of Frozen Food Producers

UK Council for Food Science and Technology

UK Federation of Business and Professional Women

Ulster Curers' Association

Ulster Farmers Union

Unilever Research

Union of Muslim Organisations of UK and Eire

Union of Orthodox Hebrew Congregations of Great Britain and the Commonwealth

United Kingdom Egg Producers' And
Retailers' Association
United Kingdom Islamic Mission
(London)
United Kingdom Islamic Mission
(Oldham)
Universities Federation for Animal
Welfare
University of Aberdeen (Faculty of
Biological Sciences)
University of Aberystwyth (Agriculture
Dept.)
University of Birmingham
University of Bristol (Dept. of Clinical
Science)
University of Durham
University of East Anglia
University of Edinburgh (Institute of
Cell, Animal and Population Biology)
University of Glasgow (Faculty of
Veterinary Medicine)
University of Greenwich
University of Newcastle-Upon-Tyne
(Agriculture and Environmental
Science)
University of Nottingham (Centre For
Applied Bioethics)
University of Reading (Centre for
Agricultural Strategy)

University of Sheffield
University of Strathclyde (Dept. of
Bioscience and Biotechnology)
University of Wales (Agricultural
Sciences)
University of York (Dept. of Biology)
UPB Porcofram Ltd.
Upjohn Ltd.
VEGA
Vegetarian Society
Veterinary Deer Society
Veterinary Science Division
Vetrepharm Ltd.
Vivash-Jones Consultants Ltd.
Waitrose Ltd.
Watson, The Revd Dr P F
Welsh Consumer Council
Western Morning News
Whittaker, Mr J
Women's Farming Union
Women's National Commission
World Rabbits Science Association
World Society for the Protection of
Animals
World Wildlife Fund for Nature
World's Poultry Science Association
(UK Branch)
Wye College (Agriculture Dept.)

ANNEX C: LIST OF CONSULTATION RESPONDENTS

Arthur Rank Centre
Association of District Councils
Babraham Institute
Baha'i Community of the United
   Kingdom
BBSRC Centre for Genome Research
Biotechnology and Biological Sciences
   Research Council (BBSRC)
British Deer Farmers Association
British Deer Society
British Goat Society
British Meat Manufacturers
   Association
British Medical Association
British Pig Association
British Poultry Meat Federation Ltd.
British Society of Animal Production
British Union for the Abolition of
   Vivisection
British Veterinary Association
Carroll, Ms A
Catholic Study Circle for Animal
   Welfare
Chartered Institute of Patent Agents
Church of Ireland Board for Social
   Responsibility (NI)
Church of Scotland, Society, Religion
   and Technology Project
Co-operative Union Ltd.
Cobb-Vantress Incorporated
Common Law Institute of Intellectual
   Property
Compassion in World Farming
Cotswold Pig Development Co. Ltd.
Council of Welsh Districts
Country Landowners' Association
Cranfield Biotechnology Centre
Deer Liaison Committee
Dixon Smith (Lyons) Ltd.
Earthkind
Farm and Food Society
Farm Animal Welfare Council
Farmers' Union of Wales
Federation of Deer Management
   Societies
Food and Drink Federation
Free Church of Scotland
Genetics Forum
Genus
Greenpeace
Harrison, Mrs Ruth
Highlands and Islands Enterprise
Holstein Friesian Society of Great
   Britain and Ireland

Imutran Ltd.
Institute of Biology
Institute of Grassland and
   Environmental Research
Kempsey, Mr R
Linzey, The Revd Professor Andrew
MacDonald, Mrs Jose
Marr, The Revd Peter
Meat and Livestock Commission
National Consumer Council
National Council of Women of Great
   Britain
National Farmers' Union
National Farmers' Union of Scotland
National Federation of Consumer
   Groups
National Federation of Women's
   Institutes
Public Health Laboratory Service
Quaker Concern for Animals
Rare Breeds Survival Trust
Roslin Institute
Royal Agricultural Society of England
Royal College of Veterinary Surgeons
Royal Society (Animal Experiments
   Committee)
Royal Society for the Prevention of
   Cruelty to Animals
Royal Society of Edinburgh
Royal Veterinary College
Royal Welsh Agricultural Society Ltd.
Scottish Agricultural College
Scottish Association of Meat
   Wholesalers
Scottish Centre for Animal Welfare
   Sciences
Scottish Society for the Prevention of
   Cruelty to Animals
Sheep Veterinary Society
Simmons and Simmons
Society for the Study of Animal
   Breeding
Universities Federation for Animal
   Welfare
University of Edinburgh (Institute of
   Cell, Animal and Population Biology)
University of Leeds
University of Nottingham (Centre for
   Applied Bioethics)
University of Strathclyde (Dept. of
   Bioscience and Biotechnology)
University of York (Dept. of Biology)
UPB Porcofram PLC

VEGA
Watson, The Revd Dr P F
Welsh Consumer Council
Women's Farming Union
Women's National Commission
World Society for the Protection of
    Animals
World's Poultry Science Association
    (UK Branch)

ANNEX D: GLOSSARY

Allele
Alternative forms of a gene which occupy the same position on a chromosome.

Artificial insemination
Artificial implantation of semen into a female animal (as opposed to insemination by natural mating).

Cell
The structural and functional unit of all living organisms. Bacteria and algae consist of one cell. Larger organisms are multicellular allowing specialization of cellular function.

Cloning
Molecular cloning is the process of replication of a single gene sequence, and may enable the production of genetically identical animals (clones).

Contained use
Any operation in which organisms are genetically modified or in which such genetically modified organisms are cultured, stored, used, transported, destroyed or disposed of, and for which physical barriers or a combination of physical barriers with chemical or biological barriers or both are used to limit their contact with the general population and the environment.

Chromosome
A large DNA (q.v.) molecular chain in the cell along which genes are located.

DNA
Deoxyribonucleic acid, which is present in all living cells and contains the information for cellular structure, organisation and function.

Embryo transfer
A procedure whereby fertilized eggs are transferred into surrogate mothers.

Epidural anaesthesia
Injection of a local anaesthetic into the epidural space of the spinal column. In posterior epidural anaesthesia the needle is inserted between the 1st and 2nd coccygeal vertebrae.

Gamete
Sperm and eggs (oocytes) are male and female gametes respectively.

Gene
The basic unit of heredity; an ordered sequence of nucleotide bases, comprising a segment of DNA. A gene contains the sequence of DNA that encodes one protein chain (via RNA q.v.). Each animal has two similar or dissimilar copies (alleles q.v.).

Genetic modification
The modification of an organism's hereditary material using artificial techniques with the aim of incorporating or deleting specific characteristics. (Also known as genetic engineering.)

Genome
The genetic endowment of an organism or individual – all of the DNA contained in a single set of chromosomes of an organism.

Genotype	Synonym for genome.
Heredity	The relation between successive generations, by which characteristics or traits are inherited.
Heterozygous	Having one or more pairs of dissimilar alleles on corresponding chromosomes, i.e. the two alternative forms of a gene for a characteristic are different.
Homozygous	Having identical rather than different alleles in corresponding positions of homologous chromosomes. The two alternative forms of a gene for a characteristic are the same and therefore the organism will breed true for that characteristic.
In vitro	Literally, in glass; pertaining to biological processes taking place in an artificial apparatus; sometimes used to include the growth of cells from multicellular organisms under cell culture conditions.
In vitro fertilization	Fertilization of an egg by sperm under laboratory conditions.
In vivo	Refers to biological processes which occur inside a living organism.
Laparoscopy	Insertion of an narrow endoscope through a small incision in the abdominal wall to view and possibly manipulate the abdominal organs. The technique may involve inflation of the abdomen with an inert gas.
Laparotomy	Surgical incision into the abdominal cavity. (Larger than that used for laparoscopy.)
Nuclear transplantation	Removal of the nucleus of one cell and transplantation into another from which the nuclear material has been removed.
Oestrus	The period of sexual receptivity to the male in the reproductive cycle of female animals.
Organism	Any biological entity, cellular or non-cellular, with capacity for self perpetuation; includes plants, animals, fungi, bacteria and viruses.
Phenotype	Appearance and behaviour of an organism resulting from the interaction between its genetic constituent and its environment.
RNA	Ribonucleic acid, which translates the information contained in genes. RNA can also be the heredity material in certain viruses.
Retrovirus	RNA viruses that utilize the enzyme reverse transcriptase during their life cycle. This enzyme allows the viral genome to be transcribed into DNA. The transcribed viral DNA is

integrated into the genome of the host cell where it replicates in unison with the genes of the host chromosome.

Selective breeding   The use of organisms exhibiting desired characteristics to produce offspring which also bear these characteristics.

Transgenic   A transgenic organism is one in which foreign genetic material has been incorporated into its genome by genetic modification.

ANNEX E: LIST OF ABBREVIATIONS USED

ACRE   Advisory Committee on Releases into the Environment

ACGM   Advisory Committee on Genetic Modification

ACNFP   Advisory Committee on Novel Foods and Processes

AI   Artificial insemination

ASPA   Animal (Scientific Procedures) Act 1986

BLUP   Best linear unbiased prediction

BST   Bovine somatotrophin

CAP   Common Agricultural Policy

EPA 90 Environmental Protection Act 1990

EPO   European Patent Office

EEA   European Economic Area

EU   European Union

FAO   Food and Agriculture Organisation of the United Nations

FAWC   Farm Animal Welfare Council

GMO   Genetically modified organism

HSE   Health and Safety Executive

IVF   *In vitro* fertilization

MAFF   Ministry of Agriculture, Fisheries and Food

RCVS   Royal College of Veterinary Surgeons

REML   Restricted maximum likelihood

UNEP   United Nations Environmental Programme

Printed in the United Kingdom for HMSO
Dd300812 2/95 C9 G3396 10170